STATISTICAL METHODS
and Financial Calculations

STATISTICAL METHODS
and Financial Calculations

ISABEL WILLEMSE

1994
JUTA & CO LTD

First published 1990
Second and third impressions 1991
Second edition 1994

© Juta & Co, Ltd
PO Box 14373, Kenwyn, 7790

This book is copyright under the Berne Convention. In terms of the Copyright Act 98 of 1978, no part of this book may be reproduced or transmitted in any form or by any means, electronic or mechanical, including photocopying, recording or by any informantion storage and retrieval system, without permission in writing from the publisher.

ISBN   0 7021 3128 8

Printed and bound by The Rustica Press, Ndabeni, Cape
D3390

Contents

1. Introduction 1
2. Collection and Organization of Data 7
3. Summarizing and Presentation of Data 17
4. Numerical Measures of Description 43
5. An Introduction to Probability 75
6. Probability Distributions 93
7. Statistical Estimation 109
8. Hypothesis Testing for Population Means and Proportions 127
9. The Chi-square Test 159
10. Analysis of Variance (ANOVA) 173
11. Simple Regression and Correlation 187
12. Time Series Analysis 201
13. Index Numbers 221
14. Statistical Decision Theory 235
15. Interest Calculations 253

APPENDICES

Revision Exercises 273
Basic Mathematics for Statistics Students 295
Statistical and Interest Tables 305

CHAPTER 1
Introduction

CONTENTS

1.1 WHAT IS STATISTICS? 1
1.2 MISUSES OF STATISTICS 2
1.3 INTRODUCTION TO BASIC TERMS 4
1.4 ROLE OF THE COMPUTER IN STATISTICS 4
1.5 STATISTICAL PROBLEM-SOLVING
    METHODOLOGY 5

1.1 WHAT IS STATISTICS?

Statistics are items of numerical information (or quantitative data). The word 'statistics' is also used, functioning as singular, to refer to the science of collecting and processing data in order to produce information.

Two usages of the word must therefore be distinguished:

(1) referring to the data themselves; and
(2) referring to the science that deals with the data.

Most people accept that an educated citizen must have an understanding of basic statistical tools to function in a modern world that is becoming increasingly dependent on qualitative and quantitative information. Statistics is not confined to our personal use only, its application is increasing also in government, business, education, medicine, science, etc. It will help you:

- make the most of data, to clarify what is inherently confusing and to study methods for making sense out of numbers and statistical language;
- describe and understand relationships that exist between variables, such as those reported in newspaper articles;
- make better decisions, for example on the quality of your product;
- cope with changing conditions; and
- provide a surer basis for decisions made under uncertainty.

Statistics is therefore concerned with the development and application of methods for the:

- collection;
- classification;
- analysis; and
- presentation of comparable numerical data,
- as well as with drawing valid conclusions and making reasonable decisions in the face of uncertainty on the basis of such data.

According to this definition the field of statistics can be subdivided into descriptive statistics and inferential statistics. Descriptive statistics includes the collection, classification, analysis and presentation of numerical data. The term 'inferential statistics' refers to the techniques of interpreting the data resulting from applying descriptive techniques to samples, and then using the results to make decisions about a population.

1.2 MISUSES OF STATISTICS

The lifeblood of management is good statistical information, and effective control of an organization's performance is based on it. A course in statistical analysis should give you the background necessary to recognise and be aware of the most commonly used methods of misrepresenting data.

Misuse of statistics is often colourful and sometimes troublesome, and it is usually done to confuse or deliberately mislead.

'There are three kinds of lies: lies, damned lies and statistics.'
Benjamin Disraeli

'He uses statistics as a drunken man uses lampposts — for support rather than illumination.'
Andrew Lang

Examples of how statistics have been misapplied

(1) The bias obstacle

The use of poorly worded or leading questions may lead to worthless conclusions. For example: 'Do you like reading fiction?' instead of: 'Which type of book do you prefer, fiction or non-fiction?' If numerical data are used and rounding is done in one direction only (for example if all answers are rounded to the next ten), an increasing error will result.

(2) Generalisation of averages and disregarded dispersions
The word 'average' can apply to several measures of central tendency such as the arithmetic mean, median or mode. It is not uncommon for one of these averages to be employed in situations in which it is not appropriate. Averages alone often do not adequately describe the true picture, because the spread of the data about the average is such that the average tends to mislead.

(3) Unwarranted cause-and-effect conclusions
Examples:
Increased shipments of peanuts into a certain port have been followed by increases in the national birthrate — therefore peanuts were the cause of the increase in births.
Ninety-five per cent of all those who smoke, consumed large amounts of milk earlier in their lives; therefore the early introduction of milk into the diet leads to smoking later in life.

(4) Failing to define terms and concepts that are important to a clear understanding of the message; making improper or illogical comparisons between unlike or unidentified items; using an alleged statistical fact or statement to jump to a conclusion; or using lengthy words to cloud the message, e.g.:
'The Trim Co. took exception to the advertised claim made by the manufacturers of Diet Bread that the latter's product had fewer calories per slice than other bread. According to the Trim Co. this claim was misleading because the Diet Bread slice on which the claim was based, was thinner than normal.'

(5) Following the bouncing base
This refers to the use of percentages as though they are comparable, although not calculated from a common base. For example:
During a recession a worker is asked to accept a 20 % reduction in his weekly wages of R300. After the recession he is given a 20 % increase. Is he happy? The answer may depend on what has happened to the base. The reduced pay amounts to R300 × 0.80 = R240. If the 20 % increase is calculated on the R240 base, the worker's 'restored' pay is still only R288.

(6) The assumption that a pattern developed in the past will continue into the future. Changes in technology, population and life-style all produce economic and social changes that may affect an existing pattern.

1.3 INTRODUCTION TO BASIC TERMS

Population: A complete collection of individuals, objects or measurements sharing a specified characteristic of interest of which the properties are to be analysed (for example: all technikon students in South Africa). The characteristics of a population can be expressed numerically in a quantitative population or non-numerically in a qualitative population. If the population contains a countable number of items, it is said to be finite, and when the number of items is unlimited, it is said to be infinite. A study of the entire population is known as a census.

Sample: A portion or part of the population of interest that is studied in order to make estimates about some unknown population characteristics.

Variable: A characteristic of interest about each individual element of a population or sample — the topic about which data are collected. Discrete and continuous variables can be distinguished by deciding whether the data result from a count or from a measurement. Example: age of people, race groups, occupation, etc. are examples of discrete variables, and mass and length of people are continuous variables.

Data (singular — datum): Observations (or an observation) collected about some topic or event. Example: a person's age would be a datum in a set of data on the variable *age of people*.

Experiment: Any process of observation or measurement.

Parameter: A numerical characteristic of a population, such as an average or standard deviation, usually indicated by a letter from the Greek alphabet.

Statistic: A numerical characteristic of a sample, usually indicated by a letter from the Roman alphabet.

Attribute: A qualitative type of description of the variable.

1.4 THE ROLE OF THE COMPUTER IN STATISTICS

The exponential growth in the volume of information has created a need for computers in business. A computer may be efficiently used in any processing operation that has one or more of the following characteristics:

(1) large volume of input;
(2) repetition of projects;

(3) desired greater speed in processing;
(4) desired and necessary greater accuracy; and
(5) processing complexities that require electronic help;

1.5 STATISTICAL PROBLEM-SOLVING METHODOLOGY

Several steps are followed in arriving at rational answers to statistical problems:

(1) Identify the problem or opportunity.
(2) Gather available facts. Data must be accurate, timely, as complete as possible, and relevant to the problem.
(3) Gather new original data if necessary.
(4) Classify and summarize the data using tables and numerical descriptive measures.
(5) Present the data using charts and diagrams.
(6) Analyse and interpret the data. If sample data were used, estimate the values of the parameters and test the assumptions about the parameters.
(7) Draw conclusions.
(8) Make decisions.

CHAPTER 2
Collection and Organization of Data

CONTENTS

2.1 SOURCES OF DATA 7
2.2 TYPES OF DATA 8
2.3 METHODS OF COLLECTION OF DATA 9
2.4 SAMPLING 12
2.5 SOURCES OF ERROR IN STATISTICAL DATA 15
2.6 DATA PREPARATION 16

The basic resource necessary for any statistical experiment is data. The quality of the final product depends on the quality of the raw material used. Researchers have adopted the acronym GIGO — garbage in, garbage out. Thus, proper data collection is extremely important to the researcher.

2.1 SOURCES OF DATA

(1) Using published or available sources of data. Internal data come from within the organization for its own use in statistical investigation, e.g. from accounting records, payrolls, inventories, salesmen, etc. External data are collected from sources outside the organization, such as trade publications, the consumer price indexes, newspapers, libraries, universities, the Central Statistical Service, etc.

(2) Designing an experiment. This provides original or raw data but often involves extremely sophisticated statistical procedures, takes a long time, and is usually very expensive.

(3) Conducting a survey.

2.2 TYPES OF DATA

Data, which are the observed outcomes of random variables, can be divided into two groups:

(1) Qualitative random variables (which yield categorical responses); and

(2) quantitative random variables (which yield numerical responses).

Each group can consist of:

❑ discrete primary or discrete secondary data; and
❑ continuous primary or continuous secondary data.

Discrete data arise from a counting process while continuous data arise from a measuring process. Both types can be obtained either internally or externally. Internal data are data obtained from within the organization (e.g. sales figures, financial statements, salaries, production, etc.) External data are obtained from outside the organization (e.g. from newspapers, magazines, libraries, the Reserve Bank, the Central Statistical Service, research organizations, etc.).

Secondary data

Available data, collected for purposes other than the current investigation, are known as 'secondary data'. Data of this type can be obtained quickly and inexpensively, but before such data can be used as the only source of information, it must be ascertained that the data are:

❑ available (has it been collected and will it be made available?);
❑ relevant (was it designed for the problem at hand?);
❑ accurate (was it obtained from an original source that can be verified and examined?); and
❑ sufficient (does it meet all requirements for the problem at hand?).

Primary data

Data specially collected for a specific purpose in order to obtain the exact information wanted, are known as 'primary data'. These data are considered more meaningful and reliable, but they are time-consuming and more costly to obtain than secondary data.

2.3 METHODS OF COLLECTION OF DATA

The main principles in data collection are:
- collect the right data, neither more nor less than are needed at the time or than will be needed in the future;
- build in sufficient checks to ensure that errors or deliberate malpractices do not go undetected.

The basis of a statistical analysis will be the data obtained in response to questions. Therefore, the questions or questionnaire must meet certain requirements.

A questionnaire can be divided into different parts:

(1) administrative part: name, date, address of respondent and interviewer, etc.;
(2) classification part: race, sex, age, marital status, occupation, etc. of the respondent; and
(3) subject matter of inquiry (questions).

Below are some of the requirements to be taken into account when drawing up a questionnaire.
- Questions must be designed to obtain the exact information as directly as possible.
- Questions should, as far as possible, not be open to misinterpretation by the respondent (ambiguous).
- Questions should be short, simple and to the point.
- The respondent must be able to understand and answer the questions.
- Questions should not require any calculations.
- Questions should be precise.
- Questions should not lead the respondent (leading questions).
- Questions should not be phrased emotively.
- Questions should not be offensive.
- The questionnaire should be well laid out, with adequate space for answers where necessary.
- Questions should follow a logical order.
- Wherever possible, a choice of answers should be given on the form, or answers should be yes, no or a tick.
- Confidentiality should be assured.

Depending on the time and money available, as well as on the sample size required and on the purpose of the study, any of a variety of methods can be used:

Personal Where the source of information is people, each may be asked a series
interviewing of questions in an interview. The information is obtained verbally.

Advantages:
- Immediate feedback.
- Cooperation — generally only about one in twenty people refuse to be interviewed in surveys.
- Interviews are fairly accurate — the interviewer can help the subject being interviewed to understand the questions.
- Difficulty with one question does not result in the loss of answers to all of them.
- The respondent does not know what other questions are coming when he answers one, and so these cannot influence his reply.
- Spontaneous answers can be obtained.
- Face-to-face contact can tell the interviewer a lot about the respondent.
- Interviews may also be conducted by telephone.

Disadvantages:
- This method is expensive because interviewers have to be trained and paid.
- The sample size is often restricted by the high costs.
- The interviewer may influence the respondent's answers.
- It may be difficult to find a convenient time for interviewing certain people.
- Respondents do not have time to consider replies or look up information.

Observation and This method entails observing the events as they happen and obtaining
inspection data by physically counting or measuring. An experiment and a traffic survey are examples of this method. Where the sources of information are objects, these must be inspected and the relevant information (e.g. size, weight, colour, etc.) must be noted.

Advantages:
- Accurate information is obtained.
- More first-hand information can be collected.
- Data can be collected over reasonably long periods of time.
- Work is not interrupted.

COLLECTION AND ORGANIZATION OF DATA

Disadvantages:
- The time and effort required push costs up.
- Considerable skill is often required.
- The past cannot be directly observed.
- It is never certain that what is observed is really representative of what happens normally, especially if people know they are being watched.
- It is not always feasible to observe the subject of the investigation (for instance how often people bath).

Abstraction Where the sources of information are documents, these must be consulted and the relevant information extracted from them. This is by far the most common method of collection in organizations; in fact, their everyday operations depend on it.

Advantages:
- Cost is low because secondary data is used.
- This may be the only method by which certain historical data can be obtained.

Disadvantages:
- Data may be unreliable.
- Data may be out of date.
- There may be certain limitations on data.
- Data may originally have been collected for a different purpose and thus not be appropriate to the current investigation.

Postal questionnaires This method is used if the targeted geographical area or number of respondents is large. A self-addressed prepaid envelope should always be included.

Advantages:
- Low cost.
- Questionnaires can be distributed widely and to many people.
- The respondent can complete the questionnaire in his own time.
- Interviewer bias is eliminated.
- Respondents have time to consider opinions, consult records, etc.
- Questionnaires can be sent out periodically to update the data.

Disadvantages:
- The response rate can be very low unless there is an incentive or a legal obligation to reply.
- There is no control over how long people take to reply.
- The answers may not be entirely the respondent's own.
- There is no spontaneity.
- Knowledge of what questions are to come later may influence answers to earlier ones.
- If a respondent has difficulty with a question, he may not return the form.
- Misunderstanding of questions cannot be corrected.
- Only straightforward questions can be set.
- Postal questionnaires are difficult to design.

2.4 SAMPLING

Sometimes only a portion of the available data is collected. This portion is known as a sample, and the process of collecting as 'sampling'. An investigation where all available data are collected, is known as a 'census'. Sampling is justified only if reasonable results can be obtained, that is, if valid inferences about the population can be drawn from the sample. If conclusions based on sampling are to be valid, sampling must be based on two general laws:

- *The Law of Statistical Regularity* holds that a reasonably large number of items selected at random from a large group of items will, on the average, have characteristics representative of the population. It is important that the selection of the sample is random so that every item in the population has an equal chance of selection. The number of items in the sample should be large enough to minimize the influence of abnormal items on the average.
- *The Law of Inertia of Large Numbers* holds that large groups of data show more stability than small ones.

Advantages:
- Costs are reduced.
- Collection time is reduced.
- Overall accuracy is improved.
- For several types of problems, sampling is the only method of data collection. An investigation may, for instance, be based on an

infinite population, or on testing procedures that entail the destruction of the item being tested (for example tests determining the life of a light bulb).

A sample is classified according to the way in which the items it comprises were chosen from the population.

Non-probability samples Any sampling technique in which the selection of sample items is not determined by chance, but rather by personal convenience, expert judgement, or any type of conscious researcher selection, is called non-probability sampling.

A major drawback of the non-probability samples is that the interviewer is given too much discretion in the process of subject selection. There is no probabilistic way of establishing to what extent a sample is representative of the population as whole.

Three principal types of non-probability sampling may be identified: convenience, judgement and quota sampling.

Convenience (chunk) sampling: These are samples made up of items selected on the basis of ready availability (convenience). It is convenient for the researcher to select the first few sample items quickly, rather than to go through the process of obtaining a probability sample. When both time and money are limited, convenience samples are widely used. The speed and ease of this sampling method ensures its existence. For example:

- Man-in-the-street interviews
- Lunch-hour interviews in the cafeteria
- Interviewing close friends or family

The errors in this type of sample are unpredictable and not measurable, and convenience samples are therefore useful only in producing ideas and not for population projections.

Judgement sampling: These samples consist of items deliberately chosen from the population on the basis of the experience and judgement of an expert in the field. Sample representatives are highly dependent upon the good judgement of the researcher. For example: a sample of four of the most influential economists was asked to estimate next year's rate of inflation. Their judgement is far superior to that of a convenience sample of the first four people found on the street, but it is not necessarily representative of the general population.

This type of sample may be subject to errors which, although relatively small, can still not be measured or predicted.

Quota sampling: In this method, the population is divided into a number of segments, and the researcher arbitrarily selects a quota of sample items from each segment. For example: Statistics students were divided on the basis of age and sex, with three age groups allocated to age, and two to sex. A certain percentage or quota to be included in each group was determined and the sample items were selected arbitrarily for each group.

	Age			
Sex	18–19	20–24	25 and over	TOTAL
Male	10 %	20 %	25 %	55 %
Female	12 %	10 %	23 %	45 %
	22 %	30 %	48 %	100 %

Probability (random) samples A probability sample is one in which the items chosen are based on known probabilities — the choice of items is left to some form of chance (random) procedure. In this type of sample the errors can be measured, and therefore the risk of making incorrect decisions can be determined.

Simple random sampling: In this type of selection every item in the population has an equal chance of selection at each successive stage of the selection process. Two of the major techniques are:

❑ the 'goldfish bowl' technique which is similar to drawing names from a hat. This method works well with a small group of people. For example: 1 000 tickets are sold to individuals who then write their names on the ticket stubs. These stubs are placed in a box and thoroughly mixed. A blindfolded person then draws the winners from the box. All 1 000 entrants have known and equal probabilities of being selected.

❑ Random numbers. Random number tables consist of rows and columns in which the decimal digits 0–9 appear. These numbers are generated by a computer programmed to scramble numbers. Any series of numbers read across, up, or down the table is considered random.

To illustrate the use of such tables, let us assume that we have 100 employees in a company and wish to interview a randomly chosen sample of ten. We could get such a random sample by assigning every employee a number from 00 to 99. Go from the top to the bottom of the columns beginning with the left-hand column, and read only the first two digits in each row. If we reach the bottom of the last column on the right and are still short of our desired number, we can go back to the beginning and start reading the third and fourth digits of each number.

According to the table of random numbers in Appendix 3, employee numbers 17, 32, 92, 40, 02, 11, 59, 31, 12 and 69 will be in the sample of 10.

Stratified sampling: This sampling method entails developing homogeneous groupings or strata across the entire population, and then taking a simple random sample within each stratum to ensure complete coverage of all types of subjects in the population. The number of items from each stratum must be proportional to the size of the stratum. For example: A researcher needs to estimate the average mass of a large group of people. He first divides the group into two strata — male and female — and then selects a simple random sample from each stratum.

Systematic sampling: A systematic sample is obtained by selecting the starting number at random and each successive number systematically from an orderly list of the population. It is used for example in quality control. Every one hundredth item leaving the assembly line is subjected to a detailed quality control inspection.

Cluster (multi-stage) sampling: This type of sample is chosen from items or people who fit a discernible description. The population is divided into clusters or strata, a few of which are randomly selected. The clusters selected are used in their entirety. For example: A large geographical area is often divided into more manageable areas or clusters. The sample is obtained by selecting a few clusters randomly from a very widespread population and then selecting a few towns randomly from each cluster or area. Out of each town select a few blocks, and out of each block select individual families at random.

2.5 SOURCES OF ERROR IN STATISTICAL DATA

The term 'error' in statistics does not mean mistake. It is defined as the difference between the 'true value' of a variable in a population and the value observed in a sample.

(1) Non-sampling errors:
- Unsystematic errors are human or mechanical errors which have no set pattern of occurrence, like misanswering of questions or mistakes in calculations. Many errors of this type can be eliminated by careful editing of returned questionnaires.

- Systematic error or bias results when there is a persistent error in one direction (over-estimation or under-estimation) during the collection process. Bias has a definite effect on the final result of the survey, and the extent of its effect can usually not be measured. A well-conducted survey should be reasonably free from bias.
(2) Random sampling error. Due to chance factors in sample selection, randomly chosen values will differ from the true population value in a random manner. This error is measurable and controllable.
(3) Survey errors. Errors of response and of non-response.

2.6 DATA PREPARATION

Once the data set has been collected, it is prepared for summary (in chart or tabular format), analysis and interpretation.

(1) Responses are scrutinized for completeness and for errors, while
(2) ensuring that there are no inconsistent entries, and
(3) entering all figures that have to be calculated.
(4) Responses to open-ended questions are properly classified and coded for data entry.

CHAPTER 3
Summarizing and Presentation of Data

CONTENTS

- 3.1 TABULATION 18
- 3.2 FREQUENCY DISTRIBUTIONS 19
- 3.3 GRAPHIC AND PICTORIAL PRESENTATION OF DATA 21
- 3.4 FREQUENCY DISTRIBUTION GRAPHS 23
- 3.5 OTHER GRAPHIC TECHNIQUES 27
- 3.6 DIAGRAMS AND CHARTS 30

This chapter discusses the arranging and displaying of raw data in a format that captures readers' attention and is easy to read and analyse. Large amounts of data will be reduced to manageable proportions. The normal procedure is to arrange data in the form of a table and then bring the figures to life by using some form of illustration.

The major principles of visual data presentation:

- ❑ Approved methods should be used so as to confirm the reliability and truthfulness of the presentation so that the reader can interpret it correctly.
- ❑ The presentation must be as simple as possible so that the reader may understand it quickly and easily.
- ❑ The presentation must interest the reader and contain worthwhile information.
- ❑ An explanation should be included in the text in order to help the reader understand the purpose of the presentation.

3.1 TABULATION

The layout of a table will depend on its purpose and the information to be presented. Most tables may be classified under the following headings:

Informative or classifying tables are original tables that contain systematically arranged data compiled for records and further use, and not intended for presentation of comparisons or to show relationships or the significance of the figures.

Reference tables contain all the summarized information relevant to the subject in question. They are usually quite long and are set out alphabetically for ease of reference.

Text or summary tables are usually found in reports, reference books, etc. They are kept as simple as possible and are usually interpreted in the accompanying text. They show only the information relevant to the question being discussed.

Basic rules for tabulation

(1) The title should be brief and should state precisely what is contained in the table.

(2) Essential additional information referring to the table as a whole or to a substantial part of it, should be included in a title note.

(3) Information which refers to a single item only should appear in a footnote.

(4) The source note explains where the data came from. It should be placed at the very bottom of the table.

(5) All rows and columns should be headed with clear, explanatory titles. The units used should be clearly defined and these, together with information about the rounding of figures, percentages, ratios, etc., should be referred to in the title note, stub or caption.

(6) Margins should be left round the edge of the table and for neatness the whole table should preferably be contained in a frame.

(7) Tables should not be overloaded. If there is a danger of this, it is often better to separate the data into more than one table.

(8) Summary statistics such as totals should be used at the end of rows and at the bottom of columns.

(9) Certain figures, such as totals, should be highlighted with double lines.

SUMMARIZING AND PRESENTATION OF DATA

3.2 FREQUENCY DISTRIBUTIONS

A frequency distribution is a table that shows how often each value or group of values occur. The data are grouped into class intervals according to some observable characteristic together with the number of observed values or frequencies that fall into each class interval.

Construction of frequency distributions

(1) Determine the range of the given ungrouped data. The range is the difference between the highest and the lowest values in the group.
(2) Determine the number of class intervals. Generally frequency distributions should contain not more than 20 classes and not fewer than 5. Sturges' rule can be used as a guideline to solve this problem.

> Sturges' Rule:
> $K = 1 + (3.3 \log n)$

(3) Determine the lengths (size or width) of each class.
Width of class = range ÷ number of classes
(4) Choose the class limits. For each class interval the stated lower and upper class limits indicate the values included within the interval.
(5) Tally the raw data into the class intervals.
(6) Count the number of tallies (observations) in each class.
(7) Construct the table.

Notes

(1) Avoid overlapping of class intervals.
(2) Class limits must be selected in such a way that both the smallest and the largest values are included. The lower and upper class limits of each interval indicate the values included within each class.
(3) The size or width of all classes should be equal.
(4) Open-ended class intervals should be avoided although they may be useful when a few values are extremely large or small in comparison with the remainder of the more concentrated values.
(5) Class intervals with a frequency of zero should be avoided.
(6) Each value can be assigned to one class interval only.
(7) The number and width of class intervals must always be whole numbers.
(8) The true class limits or class boundaries are specific points along the measurement scale which serve to separate adjoining classes and are obtained by adding the upper class limit of one class

interval to the lower class limit of next higher class interval and dividing the result by two. Class boundaries are needed to draw certain graphs and are useful in computing certain averages and measures of dispersion.

(9) The classmark or midpoint divides a class interval into two equal parts and is obtained by adding the upper and lower limits of each class interval and dividing the result by two. This value (x) represents the class interval.

(10) When the percentage of observations in each class interval instead of the actual number of observations is recorded, the distribution is known as a relative frequency distribution. This makes comparative analysis of two or more distributions relatively easy, especially when the sizes or measurement scales differ substantially. A relative frequency of a class is the observed frequency of the class divided by the total frequency of all classes.

(11) In order to determine the number or percentage of observations greater than or smaller than a particular value, a cumulative frequency distribution can be used. There are two types of cumulative frequency distribution: a 'less-than' cumulative frequency distribution is constructed by adding the frequencies from the lowest to the highest class interval. A 'more-than' cumulative frequency distribution is constructed by adding the frequencies from the highest to the lowest class interval.

Example A firm wants to analyse the characteristics of its employees. A sample of 120 employees is selected, and the age at nearest birthday of each is determined. The following table shows the ages as recorded:

60	39	23	30	29	26	29	41	40	32	63	22
32	52	46	35	25	28	33	33	20	25	42	34
29	43	41	31	30	36	58	21	24	55	51	28
18	40	44	38	32	21	30	31	25	49	31	26
33	36	43	34	35	22	33	38	34	34	33	34
23	26	57	23	26	36	39	31	35	34	34	51
40	50	35	45	28	36	32	39	26	48	17	45
45	25	25	30	36	30	43	25	27	21	53	25
38	33	37	33	18	24	30	36	44	49	54	60
19	25	26	33	35	43	56	29	48	31	45	24

Range

$R = 63 - 17 = 46$

SUMMARIZING AND PRESENTATION OF DATA

Number of intervals
$K = 1 + 3.3 \log n$
$K = 1 + 3.3 \log 120$
$K = 7.86 \approx 8$

Width of interval (c)
$c = R \div K = 46 \div 8 = 5.75 \approx 6$

Class limits		f
17–22:	‖‖‖‖ ‖‖‖‖	= 10
23–28:	‖‖‖‖ ‖‖‖‖ ‖‖‖‖ ‖‖‖‖ 1111	= 24
29–34:	‖‖‖‖ ‖‖‖‖ ‖‖‖‖ ‖‖‖‖ ‖‖‖‖ ‖‖‖‖ 1111	= 34
35–40:	‖‖‖‖ ‖‖‖‖ ‖‖‖‖ ‖‖‖‖ 1	= 21
41–46:	‖‖‖‖ ‖‖‖‖ 1111	= 14
47–52:	‖‖‖‖ 111	= 8
53–58:	‖‖‖‖ 1	= 6
59–64:	111	= 3
TOTAL		120

Class boundaries	f	%f	cum <f	cum <%f	cum >f	cum >%f	x
16.5–22.5	10	8	10	8	120	100	19.5
22.5–28.5	24	20	34	28	110	92	25.5
28.5–34.5	34	28	68	56	86	72	31.5
34.5–40.5	21	18	89	74	52	44	37.5
40.5–46.5	14	12	103	86	31	26	43.5
46.5–52.5	8	7	111	93	17	14	49.5
52.5–58.5	6	5	117	98	9	7	55.5
58.5–64.5	3	2	120	100	3	2	61.5
TOTAL:	120	100					

3.3 GRAPHIC AND PICTORIAL PRESENTATION OF DATA

A graph does not replace a table, but complements it by showing the data's general structure more clearly. It is more likely to receive the attention of the casual observer and reveal trends or relationships that might be overlooked in a table. Many types of pictorial presentation forms are employed in statistics according to the nature and purpose of the data. A graph shows the relationship between two variables, or changes in a variable over a time period. One of the variables will be independent (x) and the other dependent (y).

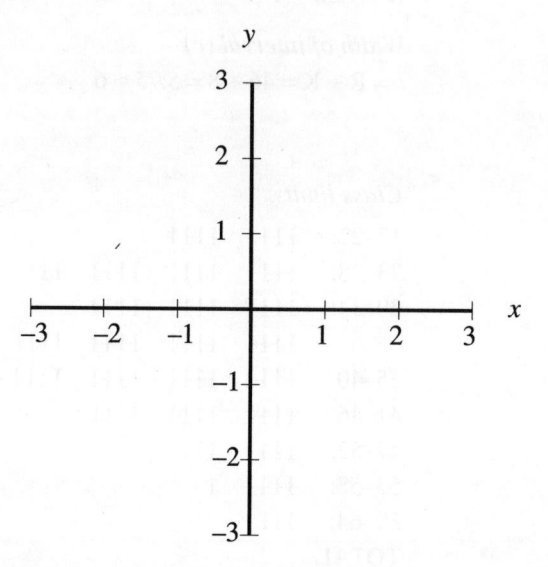

Notes: (1) A graph is drawn on a grid to a certain scale by drawing two lines at right angles, known as coordinate axes. The point of intersection of the two axes is called the origin (0).
(2) Title notes, titles, footnotes, etc. are used as for tables.
(3) Scale captions should show units, ratios, percentages, etc. and are usually expressed as multiples of 5, 10, 100 etc.
(4) Lettering should be horizontal.
(5) The scales should be shown at the left of the upright axis and below the horizontal axis.

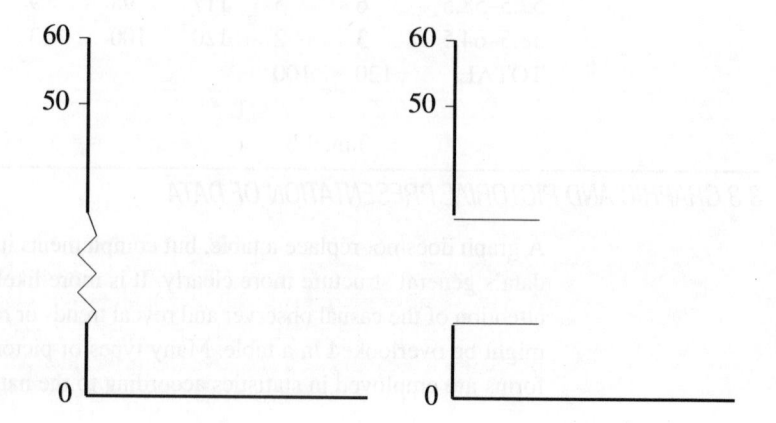

(6) The graph should be clearly distinguished from the coordinate lines.
(7) If more than one curve occurs on the same axis, the curves must be clearly distinguished.
(8) The zero line on the vertical scale should, if possible, always be shown. If this is not done, fluctuations may be over-emphasized. If it is not possible to show a complete scale, the methods shown at the bottom of page 22 should be used.

Common methods of presenting statistical data:
- Frequency distribution graphs
- Line graph techniques
- Charts and diagrams

3.4 FREQUENCY DISTRIBUTION GRAPHS

Graphs of frequency distributions are useful because they accentuate and clarify trends that are not as readily observable in tables.

A frequency table may be graphed in three ways:
(1) The histogram and relative histogram
(2) The polygon and relative polygon
(3) The cumulative curve (ogive) and relative ogive

The histogram A histogram is a continuous series of rectangles of equal width but different heights drawn to display the class frequencies.

- Points corresponding to the class boundaries of the variable being studied are marked on the x-axis. If the class intervals are equal in size, the points must be equidistant from one another.
- On the y-axis the frequencies of the class intervals are marked. A proper scale showing the true zero must be used on the y-axis in order not to misrepresent the character of the data.
- Rectangles are drawn with widths equal to the class intervals and heights determined by the corresponding class frequencies. The area of each rectangle is proportional to the frequency of the class interval it represents. The total area of the rectangles represents the total number of observations or "total of the frequencies".

A relative histogram is a graph that displays the class boundaries of the intervals on the x-axis and the relative frequencies (%f) of the intervals on the y-axis. The relative frequency histogram of a particular data set has a similar shape to the frequency histogram for that set, because in both instances the relative size of each rectangle is the frequency of that interval compared with the total number of observations.

Example: The record of daily sales of a small retail shop during a recent period is summarized in the following frequency distribution.

Daily Sales	Class boundaries	Number of days	% of days	x
90–99	89.5–99.5	18	24	94.5
100–109	99.5–109.5	25	33	104.5
110–119	109.5–119.5	12	16	114.5
120–129	119.5–129.5	9	12	124.5
130–139	129.5–139.5	6	8	134.5
140–149	139.5–149.5	3	4	144.5
150–159	149.5–159.5	2	3	154.5
TOTAL:		75	100	

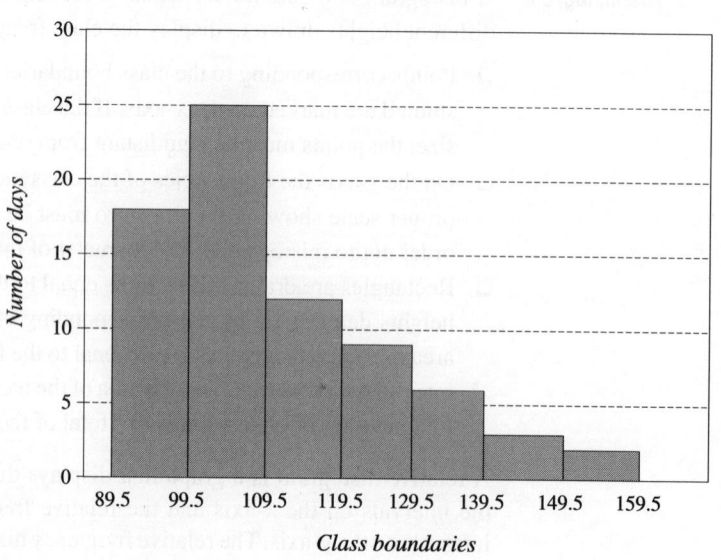

HISTOGRAM
Daily Sales

SUMMARIZING AND PRESENTATION OF DATA

Frequency polygons The frequency polygon is a line graph which can also be used to portray the shape of the distribution. It is used less widely than the histogram, because it does not show the relation of each class frequency to the total as clearly as the histogram does. To construct the polygon, the frequency distribution must be modified in a different manner:

- Find the midpoint (x) of each class interval and assume that each class is represented by its midpoint.
- Mark the frequencies on the y-axis using a proper scale and preferably starting at the true zero point. The scale must include values large enough to include the largest class frequency.
- Mark the class midpoints on the x-axis.
- Plot each class frequency by drawing a dot above its corresponding midpoint at a height equal to the frequency.
- Connect the successive dots with a straight line to form the polygon.

Polygon of daily sales (refer previous example)

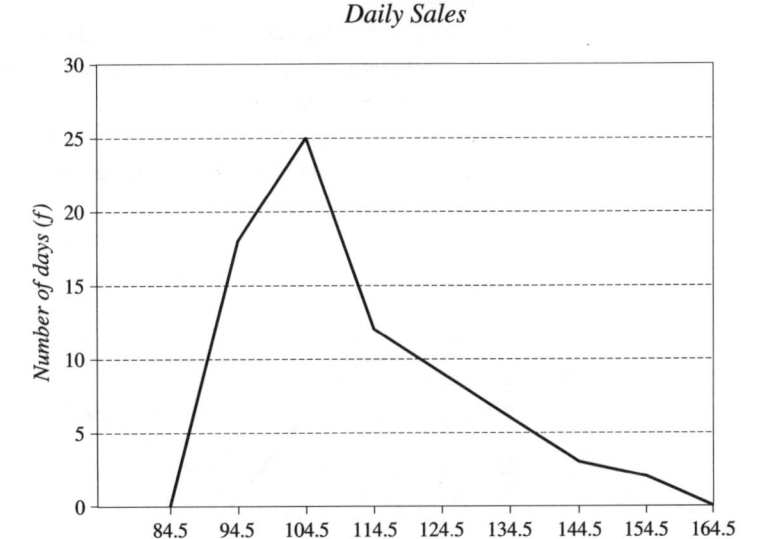

POLYGON
Daily Sales

[Frequency polygon chart with x-axis "Class midpoints (x)" showing values 84.5, 94.5, 104.5, 114.5, 124.5, 134.5, 144.5, 154.5, 164.5 and y-axis "Number of days (f)" ranging 0 to 30, peaking at 25 around midpoint 114.5]

- It is customary to close the polygon at both ends by adding classes with zero frequencies at each end of the observed midpoints on the x-axis. The two new points allow the polygon to touch the x-axis at both ends.

If the class intervals in a frequency distribution were continuously reduced in size, and if the number of observations were continuously increased, we would expect the polygon to resemble a smooth curve known as a frequency curve.

A polygon that uses the relative frequencies of the intervals rather than the actual number of points is called a relative frequency polygon. It has the same shape as the frequency polygon, but uses a percentage scale on the y-axis.

Cumulative frequency curves (ogive) A cumulative frequency curve is a smooth curve that can be used to estimate graphically the number of items with values less than some specified value (or greater, if the ogive is 'more than'), rather than merely recording the number of items within intervals.

- As the name implies, an ogive requires cumulative class frequencies. The 'less-than' ogive requires cumulative 'less-than' frequencies and the 'more-than' ogive requires cumulative 'more-than' frequencies.
- The frequency scale on the y-axis must extend to the total of the frequencies.
- The class boundaries are laid out on the x-axis starting with the lower boundary of the first class, and ending with the upper boundary of the last class.
- For the cumulative 'less-than' ogive the cumulative 'less-than' frequencies are plotted together with the upper boundary of each class.

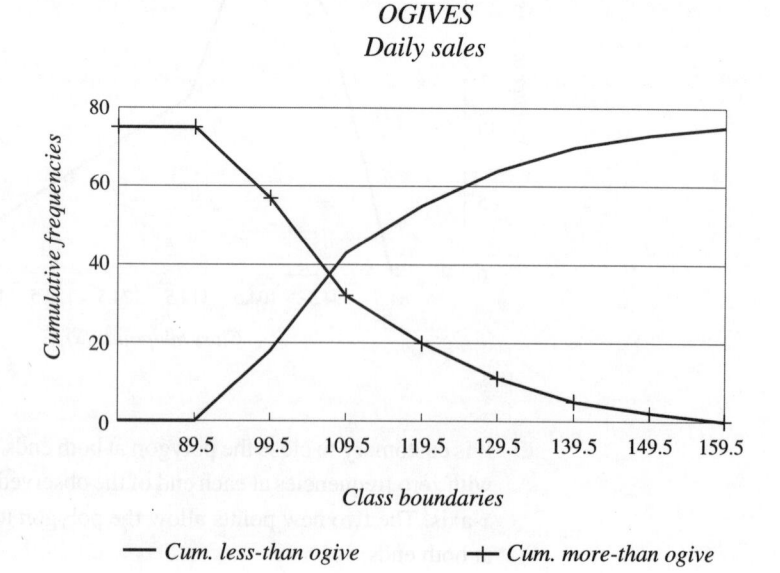

OGIVES
Daily sales

——— *Cum. less-than ogive*   —+— *Cum. more-than ogive*

SUMMARIZING AND PRESENTATION OF DATA

- For the cumulative 'more-than' ogive the cumulative 'more-than' frequencies are plotted at the lower boundary of each class.
- Draw a smooth curve through the points. The 'less than' curve slopes upward and to the right. The 'more than' curve slopes downward and to the right.
- The 'less than' ogive starts on the left with a frequency of zero at the lower class boundary of the first class.
- The 'more than' ogive ends at the upper class boundary of the last class with a frequency of zero.

If the cumulative frequencies are expressed as percentages of the total, a relative ogive can be drawn.

Example
(Using the same figures as the previous example — Daily Sales)

Class bounda-ries	Number of days	% of days	cum < f	cum > f
89.5 –99.5	18	24	18	75
99.5–109.5	25	33	43	57
109.5–119.5	12	16	55	32
119.5–129.5	9	12	64	20
129.5–139.5	6	8	70	11
139.5–149.5	3	4	73	5
149.5–159.5	2	3	75	2
TOTAL	75	100		

3.5 OTHER GRAPHIC TECHNIQUES

Line Chart A line chart is particularly useful for demonstrating changes in data over a period of time or to show the relationship that exists between two variables.

Percentage line charts can be used to show the changes of one variable relative to another over a period of time, and to emphasize fluctuations about the 100 % line.

Lorenz curve This curve is a graphic method of showing to what extent different variables deviate from uniformity, or to show degrees of deviation in order to make comparisons. If two variables are constant relative to each other, a percentage change in one variable will result in the same percentage change for the other variable. This situation may be repre-

Example:

LINE CHART

MOTORCYCLE REGISTRATIONS IN SOUTH AFRICA
No. of Registrations in Millions

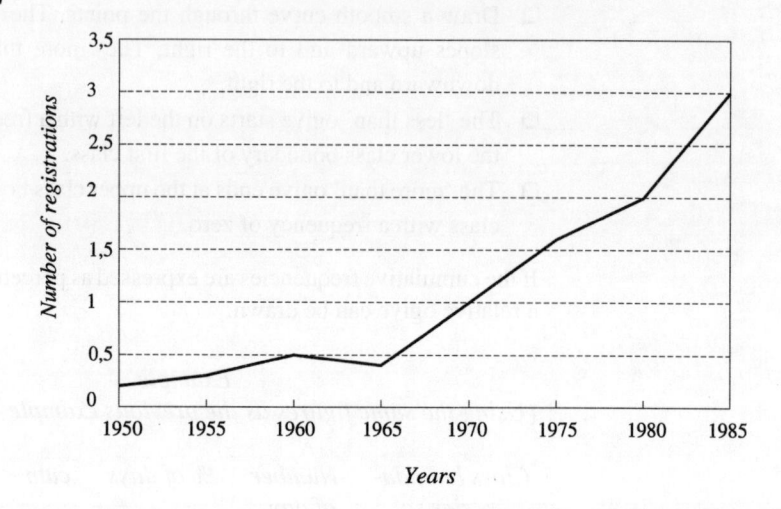

[Line chart showing number of registrations (0 to 3.5 millions) on y-axis versus Years (1950 to 1985) on x-axis. The line starts near 0.2 in 1950, rises slowly to about 0.5 by 1960, dips slightly by 1965, then rises steeply to about 3 by 1985.]

Years

sented on a graph by a straight line. In the case of the Lorenz curve, this line is always at 45° to the origin and is known as the line of uniformity or equal distribution.

❏ Draw up a frequency table with cum. 'less-than' percentage columns for both variables.

❏ Plot the cum. percentages against each other on the two percentage axes.

❏ Draw the 45° diagonal.

Upper limit of annual savings	No of savers (f)	Amount saved	cum. %f	cum.% amount
0.50	9 133	2 511	80.4	29.5
1.00	1 503	998	93.7	41.3
5.00	546	1 002	98.5	53.0
15.00	95	971	99.3	64.5
35.00	51	1 024	99.8	76.5
100.00	22	985	100.0	88.1
100.00 and more	5	1 013	100.0	100.0
TOTAL	11 355	8 504		

SUMMARIZING AND PRESENTATION OF DATA

LORENZ CURVE
Annual Savings

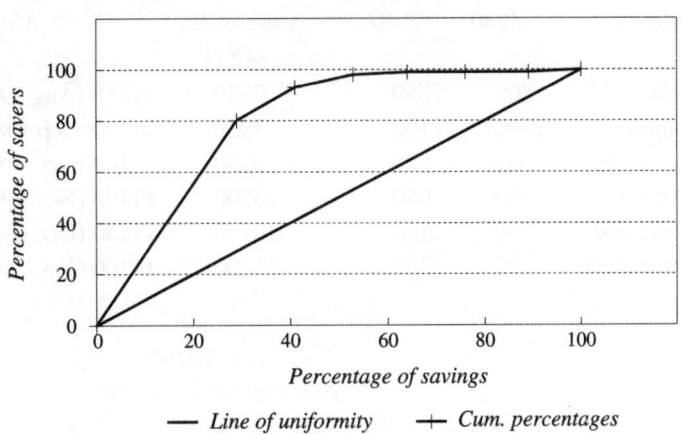

[Chart: Lorenz curve with Percentage of savers (y-axis, 0-100) vs Percentage of savings (x-axis, 0-100), showing Line of uniformity and Cum. percentages]

Source: Annual Report of Rich Bank

Z-curve   This chart combines three different curves. The name 'Z-curve' refers to the chart's resemblance to the letter Z. Z-charts may be used for any data where short-term movements in relation to long-term is of interest. Each Z cycle normally takes one year. The three curves are in respect of:

- the original monthly data (showing the seasonal fluctuations);
- cumulative original monthly data (showing the position up to date and whether the sales have increased steadily over the year or not); and
- moving annual total (the total figure for the twelve preceding months, showing the long-term trend).

SALES OF ABC CO.

Month	1980	1981	Cum. sales 1981	Moving total
January	350	400	400	11150 (Feb 1980 to Jan 1981)
February	350	450	850	11250 (Mar 1980 to Feb 1981)
March	800	800	1650	11250 (Apr 1980 to Mar 1981)
April	1000	950	2600	11200 (May 1980 to Apr 1981)
May	1550	1500	4100	11150 (Jun 1980 to May 1981)
June	1850	1900	6000	11200 (Jul 1980 to Jun 1981)

SALES OF ABC CO.

Month	1980	1981	Cum. sales 1981	Moving total
July	1800	1850	7850	11250 (Aug 1980 to Jul 1981)
August	1200	1100	8950	11150 (Sep 1980 to Aug 1981)
September	650	600	9550	11100 (Oct 1980 to Sep 1981)
October	650	650	10200	11100 (Nov 1980 to Oct 1981)
November	500	550	10750	11150 (Dec 1980 to Nov 1981)
December	400	350	11100	11100 (Jan 1981 to Dec 1981)

Z-CHART
Sales of ABC Company

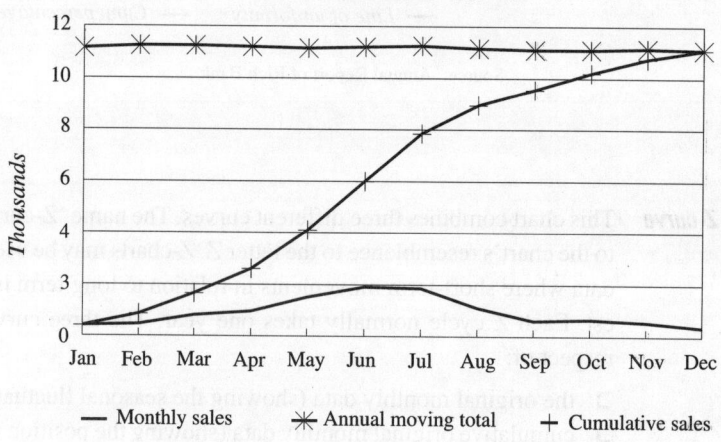

— Monthly sales   ✳ Annual moving total   + Cumulative sales

Source: Financial statements 1981

3.6 DIAGRAMS AND CHARTS

The three main methods of diagrammatic presentation are:

- bar charts;
- pie charts; and
- pictographs.

Bar charts Rectangles of equal width are drawn so that the area enclosed by each rectangle is proportional to the size of the variable it represents. This type of graph not only illustrates a general trend, but also allows a quick and accurate comparison of one period with another.

SUMMARIZING AND PRESENTATION OF DATA

Simple bar chart Single bars representing each variable are drawn either vertically or horizontally.

Simple bar chart

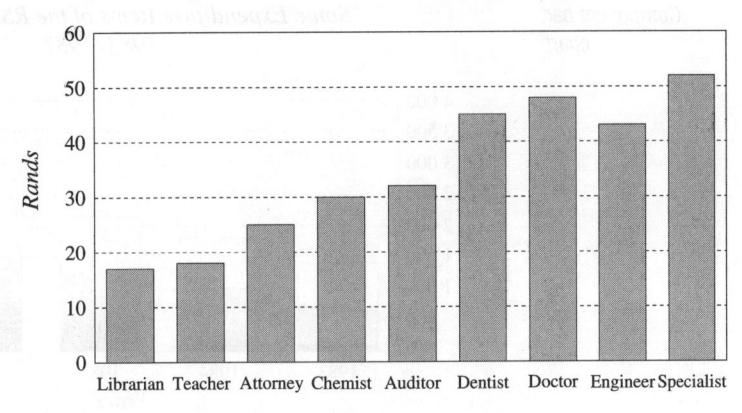

**Median Income
According to Occupation 1980**

[Bar chart showing Rands on y-axis (0-60) for occupations: Librarian (~17), Teacher (~18), Attorney (~25), Chemist (~30), Auditor (~32), Dentist (~45), Doctor (~48), Engineer (~43), Specialist (~52)]

Source: *Finance News,* November 1980

Multiple (compound) bar chart This type of chart shows several variables over the same time period, or a given variable over several periods. Two or more bars are grouped together and more than one set of comparisons can be made. The use of a key will help distinguish between the categories.

Multiple bar chart

**Growth Rate of Government Expenditure
1978–1986**

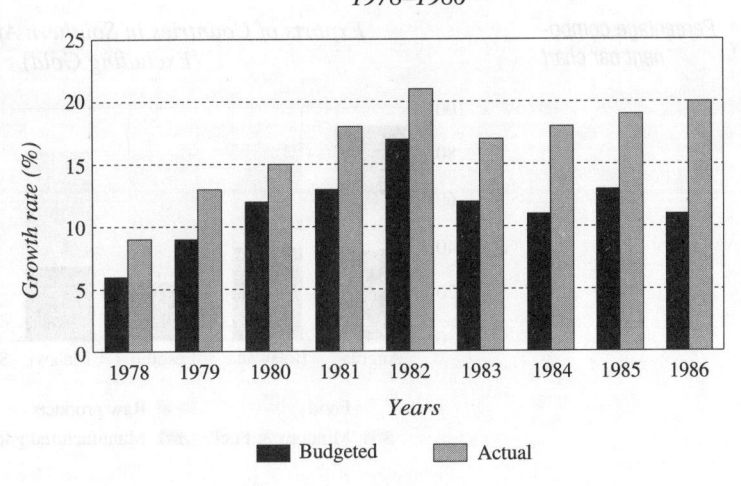

[Bar chart showing Growth rate (%) on y-axis (0-25) for years 1978-1986, with Budgeted and Actual bars for each year]

■ Budgeted   ▨ Actual

Source: *Focus on Key Economic Issues*, No. 35.

STATISTICAL METHODS AND FINANCIAL CALCULATIONS

Component or sub-divided bar chart

Each bar is divided into sections that represent the components that make up the total bar. Descriptions of the items involved may appear on the sections of the bar, or they may be identified by colouring or shading, accompanied by an explanatory key.

Component bar chart

Some Expenditure Items of the RSA Government 1983–1987

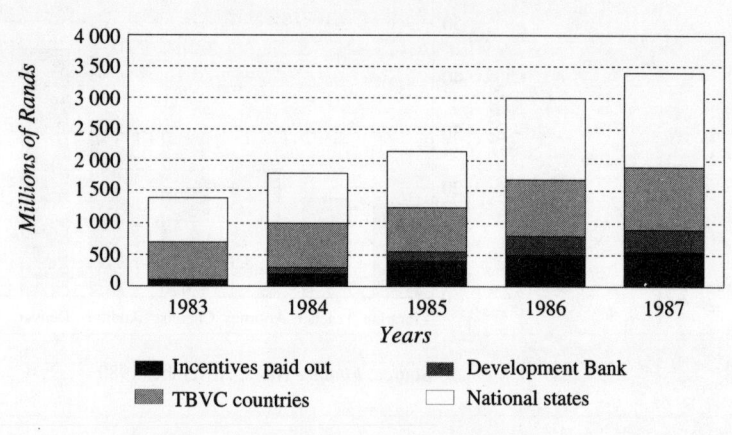

■ Incentives paid out   ■ Development Bank
■ TBVC countries   □ National states

Source: *Focus on Key Economic Issues*, No. 39

Percentage component bar chart

The components are converted to percentages of the total, and the bars are divided in proportion to these percentages.

The scale is a percentage scale and all bars must be of the same length (100 %). A key will also be necessary.

Percentage component bar chart

Exports of Countries in Southern Africa for 1986 (Excluding Gold)

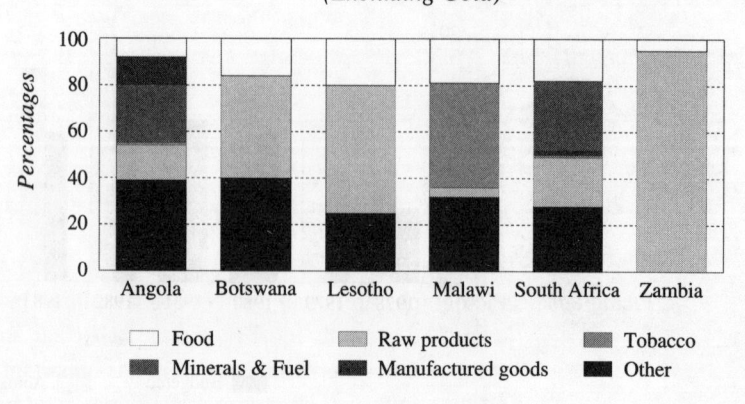

□ Food   ■ Raw products   ■ Tobacco
■ Minerals & Fuel   ■ Manufactured goods   ■ Other

Source: *Focus on Key Economic Issues*, No. 12

Pie charts A pie chart represents information in the form of a circle divided into segments showing the relationship of parts to the whole.

Steps:

- Rank the data by size.
- Calculate each value as a percentage of the total.
- Multiply each percentage by 360° to determine the segment size.
- Draw the segments with the help of a protractor, starting with the largest. Label each segment with the percentage of the total it represents.

Exports by Product Group
1980–1984 Average

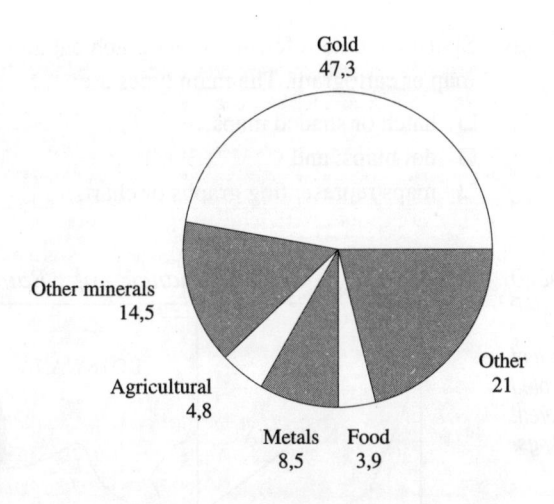

Gold 47,3
Other minerals 14,5
Agricultural 4,8
Metals 8,5
Food 3,9
Other 21

Source: *Focus on Key Economic Issues*, No. 3

Pictographs Pictographs are small symbols or simplified pictures that represent data. The symbols must be simple and clear and the quantity each symbol represents should be given. It is important that the symbols are all the same size, otherwise readers may interpret the differences in size as reflecting area or volume.

Example of a pictograph

Acme Van Manufacturing Company
Output of Vans (1 unit = 1 000 vans)

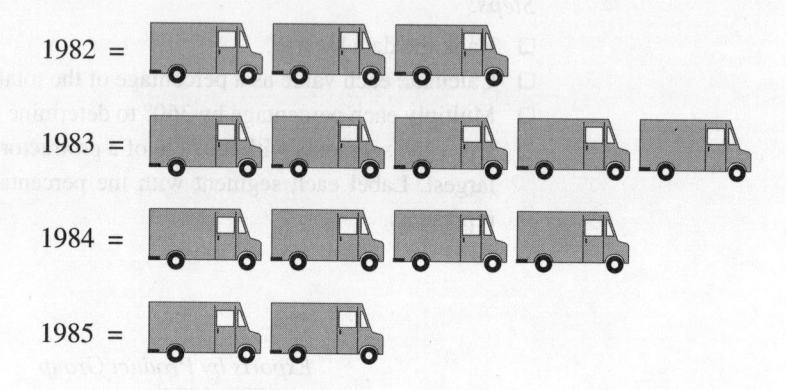

1982 = 🚐 🚐 🚐

1983 = 🚐 🚐 🚐 🚐 🚐

1984 = 🚐 🚐 🚐 🚐

1985 = 🚐 🚐

Statistical maps Statistical data referring to geographical areas can be presented as a map or cartogram. The main types are:

❑ hatch or shaded maps;
❑ dot maps; and
❑ maps representing graphs or charts.

HATCH OR SHADED MAP

The varying size of the data is denoted by different shadings.

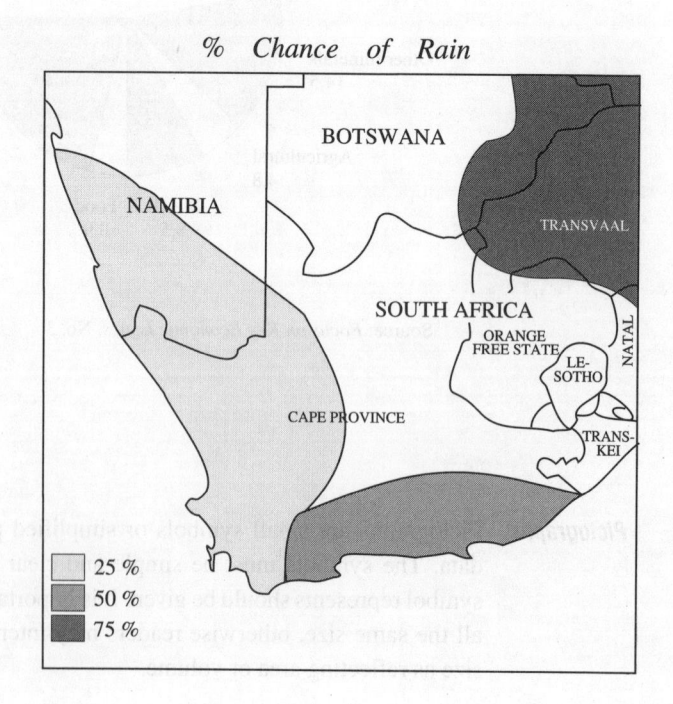

% Chance of Rain

(Map showing Namibia, Botswana, South Africa, Transvaal, Orange Free State, Lesotho, Natal, Cape Province, Transkei)

◻ 25 %
▩ 50 %
▪ 75 %

SUMMARIZING AND PRESENTATION OF DATA   35

DOT MAP

The size of a dot reflects the size of the number it represents.

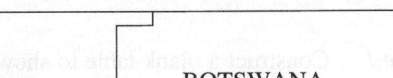

XYZ's SALESPEOPLE IN SA

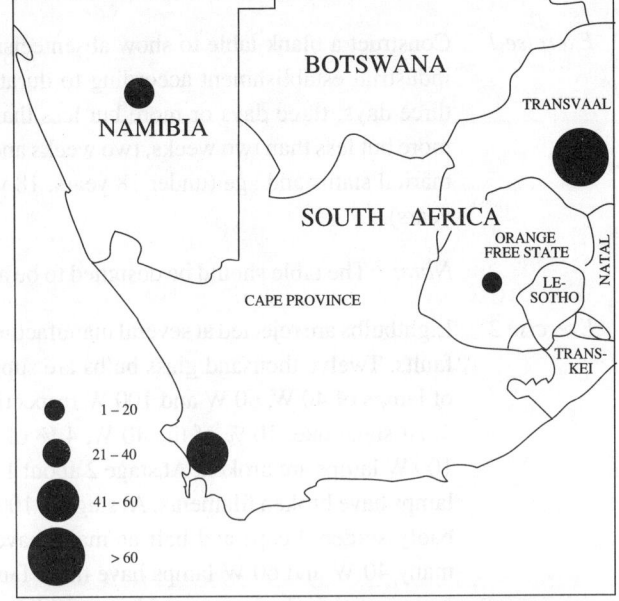

- 1 – 20
- 21 – 40
- 41 – 60
- > 60

MAPS REPRESENTING GRAPHS OR CHARTS

Daily temperature forecast

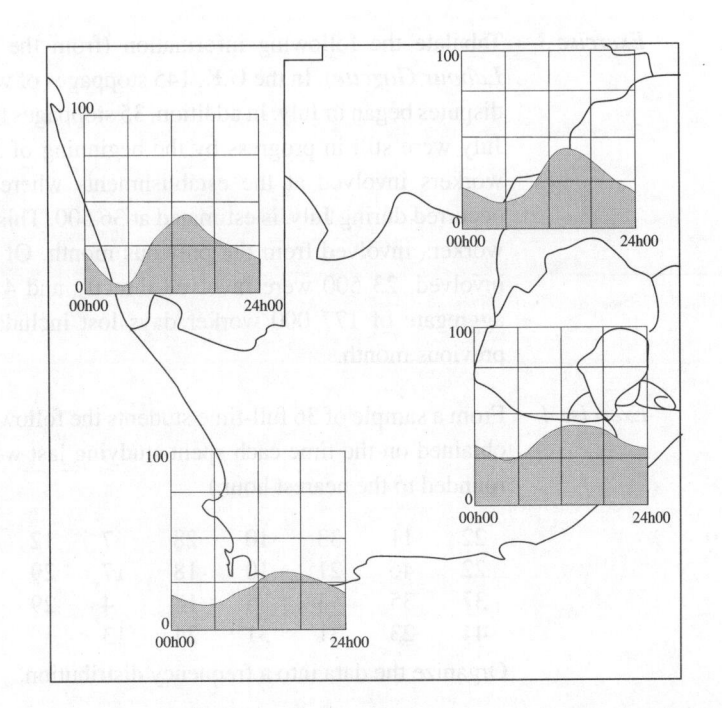

EXERCISES

Exercise 1 Construct a blank table to show absenteeism of the employees of an industrial establishment according to duration of absence (less than three days, three days or more but less than one week, one week or more but less than two weeks, two weeks and longer), sex of absentee, marital status and age (under 18 years, 18 years to under 50, over 50 years).

Note: The table should be designed to be applied to one month only.

Exercise 2 Lightbulbs are rejected at several manufacturing stages due to different faults. Twelve thousand glass bulbs are supplied for the manufacture of lamps of 40 W, 60 W and 100 W respectively, in the ratio of 1 : 2 : 3. At stage one, 10 % of the 40 W, 4 % of the 60 W and 5 % of the 100 W lamps are broken. At stage 2 about 1 % of the remainder of the lamps have broken filaments. At stage 3, 100 of the 100 W lamps have badly soldered caps and half as many have crooked caps. Twice as many 40 W and 60 W lamps have these faults. At stage 4 about 3 % are rejected for bad type-marking and 1 in every 100 are broken in the packing which follows. Arrange this information in a tabular form. Which type of lamp shows the greatest wastage during manufacture? (Source: Institute of Statisticians)

Exercise 3 Tabulate the following information (from the British *Ministry of Labour Gazette*). In the U.K. 145 stoppages of work due to industrial disputes began in July. In addition, 35 stoppages that had begun before July were still in progress by the beginning of July. The number of workers involved at the establishments where the 180 stoppages occurred during July, is estimated at 36 500. This total includes 8 400 workers involved from the previous month. Of the 28 100 workers involved, 23 500 were involved directly and 4 600 indirectly. The aggregate of 177 000 worker days lost included 94 000 from the previous month.

Exercise 4 From a sample of 36 full-time students the following information was obtained on the time each spent studying last week. (Time is shown rounded to the nearest hour.)

22	11	33	10	28	7	12	23	21	2
22	46	21	10	18	17	29	25	10	20
37	35	3	5	18	4	29	14	32	36
44	23	31	31	24	13				

Organize the data into a frequency distribution.

SUMMARIZING AND PRESENTATION OF DATA

Exercise 5   A sample of the homes sold during the past year by the Best Home Company was selected. (Selling price is in thousands of rands.)

75.6	93.9	70.7	97.9	79.6	77.9	79.9	81.7	67.2	87.9
74.8	98.4	85.3	85.0	86.6	82.0	98.2	80.0	72.3	92.3
89.7	94.6	94.0	77.6	90.7	70.4	81.7	85.7	93.0	76.9

Organize the data into a frequency distribution.

Exercise 6   The following figures are the number of kilometers (in thousands, and rounded to the nearest 1 000) driven during the year by 110 salesmen. Prepare a frequency distribution, a cumulative frequency distribution, a histogram, a relative histogram, a polygon, a cumulative 'more-than' and a cumulative 'less-than' ogive for these data.

40	26	41	40	39	34	61	42	47	23	18
43	29	93	46	32	44	71	45	62	36	22
49	31	35	36	84	81	51	51	52	66	34
55	44	18	33	38	28	42	11	48	55	42
65	54	97	67	88	44	39	42	35	50	90
73	60	41	40	29	24	58	47	53	45	84
30	31	32	34	48	76	38	52	63	41	73
36	50	31	56	35	15	26	28	41	45	61
32	27	75	30	68	24	37	30	20	50	52
10	65	52	20	36	38	38	43	21	55	48

Exercise 7   The following is an ordered array of the amounts of money (in millions of rands) on deposits in each of 100 banks on a certain date. From these data construct the following: a frequency distribution, a cumulative frequency distribution, a relative frequency distribution, a cumulative relative frequency distribution, a histogram and an ogive.

0.9	2.3	5.0	6.0	8.6	10.3	13.7	16.1	21.3	27.5
1.1	2.4	5.1	6.1	8.7	10.5	13.9	16.3	21.2	28.3
1.5	2.5	5.2	6.1	8.8	11.1	14.0	16.4	22.4	28.3
1.7	2.7	5.2	6.5	9.3	11.2	14.2	17.1	23.6	29.4
1.8	3.0	5.4	6.8	9.4	11.8	14.4	17.2	23.8	29.0
1.9	3.2	5.5	6.9	9.5	12.1	14.5	18.8	24.0	30.0
1.9	3.7	5.6	7.1	9.6	12.2	14.6	19.0	24.2	30.1
2.0	4.2	5.7	7.3	9.8	13.5	15.1	19.5	25.4	30.5
2.0	4.6	5.8	7.4	9.8	13.6	15.3	20.4	25.2	33.2
2.1	4.9	5.8	8.2	10.1	13.6	15.6	20.5	26.2	34.4

Exercise 8 The age distribution of guards hired by the Jumbo Department within the last year is:

Age	Number
20–29	2
30–39	13
40–49	20
50–59	12
60–69	3

(a) Determine the class boundaries of the distribution.

(b) Construct a histogram and polygon of these data.

(c) Construct a 'less-than' and 'more-than' cumulative frequency distribution.

(d) Draw the ogives.

(e) Based on the graph, the age of about half the guards hired was less than . . . .

(f) Based on the graph, the age of about 30 of the guards hired was more than . . . .

Exercise 9 For the following frequency distribution construct:

(a) a relative histogram and relative polygon; and

(b) A % 'less-than' and a % 'more-than' ogive.

(c) What percentage of the frequencies is more than 3.595?

(d) What percentage of the frequencies is less than 3.995?

3.00–3.19	1
3.20–3.39	4
3.40–3.59	11
3.60–3.79	15
3.80–3.99	12
4.00–4.19	11
4.20–4.39	8
4.40–4.59	7
4.60–4.79	6

Exercise 10 The following relative frequency distribution resulted from a study of the rand amount spent per visit by customers at a supermarket.

Amount spent	%f
1.00–5.99	1
6.00–10.99	3
11.00–15.99	4
16.00–20.99	6
21.00–25.99	7
26.00–30.99	9
31.00–35.99	11
36.00–40.99	19
41.00–45.99	32
46.00 and above	8

(a) Determine the class midpoints and class boundaries for each interval.
(b) If the study covered a sample of 3 000 customers, how many of the customers fall into each class interval.

Exercise 11 The total foreign investment in South Africa (in billions of rands) between 1975 and 1981 is shown in the table that follows:

Year:	1975	1976	1977	1978	1979	1980	1981
Investment:	174.5	197.4	220.5	263.4	300.9	307.8	370.4

(a) Portray the foreign investment as a line chart.
(b) Portray the foreign investment as a simple bar chart.

Exercise 12 Illustrate the following data by means of a bar chart:

Johnson and Company: Analysis of Costs

Year	Wages (R'000)	Raw materials (R'000)	Overheads (R'000)
1979	10	9	3
1980	12	10	3
1981	15	12	3.5
1982	14	11	4.5
1983	10	8	3
1984	11	8	2
1985	12	9	2

Exercise 13 Use (*a*) a multiple bar chart, and (*b*) a sub-divided bar chart to illustrate the following data:

Comparisons of sales for subsidiaries of A.Alvis Co.

Year	City	Sales
1979	East London	2
	Johannesburg	18
	Cape Town	14
1980	East London	3
	Johannesburg	26
	Cape Town	22
1981	East London	5
	Johannesburg	33
	Cape Town	32
1982	East London	2
	Johannesburg	16
	Cape Town	16

Exercise 14 The following figures came from a report on the production census:

Textile Machinery and Accessories.

No. of machines	Net output (R)
48	1 406
42	2 263
38	3 699
21	2 836
26	3 152
16	5 032
23	8 385

Analyse this table by means of a Lorenz curve and explain what this curve shows.

Exercise 15 From the sales of Lino Products Ltd. given below, draw a Z-chart for 1994.

	1993	1994
January	143	156
February	172	182
March	181	201
April	209	234
May	412	512

SUMMARIZING AND PRESENTATION OF DATA

	1993	*1994*
June	413	511
July	208	306
August	135	214
September	118	239
October	241	337
November	326	385
December	389	411

Exercise 16   From the following data, plot a Z-chart for 1995.

	1994	*1995*
January	450	550
February	450	600
March	900	850
April	1 100	1 150
May	1 600	1 160
June	1 900	1 950
July	1 850	1 900
August	1 350	1 400
September	700	850
October	700	800
November	650	700
December	550	600

Exercise 17   Use a percentage chart to show the relationship of net profit to sales.

Year	*Net Profit (R)*	*Sales (R)*
1978	800	4 000
1979	850	4 200
1980	950	4 200
1981	1 000	4 000
1982	1 100	4 400
1983	1 200	4 200
1984	1 250	4 400
1985	1 400	4 800

Exercise 18 Use a pie chart to illustrate the following data:

Waxman & Co
Analysis of employee hours lost

Hours lost through illness	5 100
Hours lost through holidays	2 150
Hours lost through industrial disputes	8 750
Total hours lost	16 000

CHAPTER 4
Numerical Measures of Description

CONTENTS

4.1 MEASURES OF CENTRAL TENDENCY 44
4.2 MEASURES OF NON-CENTRAL POSITION 55
4.3 MEASURES OF DISPERSION 57
4.4 MEASURES OF RELATIVE DISPERSION 63
4.5 MEASURES OF SHAPE 64

The major drawback of making inferences about a population from graphical descriptions based on sample data is that the reliability of the inference cannot be measured. To provide such a measure, some numerical description of the characteristics of a data set is needed. These single numbers or summary statistics that are either typical of the data set or best describe it, will make quicker and better decisions possible because they eliminate the need to consult original observations.

The most important summary statistics that have been developed are:

(1) Measures of central tendency.
(2) Measures of dispersion.
(3) Measures of shape.

4.1 MEASURES OF CENTRAL TENDENCY

These measures indicate typical or middle data points or typical representative figures. In most cases the data tend to cluster in the middle around some value which is often referred to as an average. There are many averages, each having distinct characteristics, advantages and disadvantages. For the same set of data all the averages might have different values. Three commonly used averages are:

- the arithmetic mean;
- the median; and
- the mode.

The arithmetic mean This is the measure of central tendency most commonly used. To determine the arithmetic mean, the sum of the values is obtained and divided by the number of observations.

Ungrouped data Ungrouped data are data that have not been put into a frequency distribution. Original data points are used in calculations.

$$\bar{x} = \frac{\sum x}{n}$$

Where:

$\bar{x}$ = arithmetic mean

$\sum$ = sum of

x = each individual value

n = number of values

Steps:

- Add the values of the individual numbers ($\sum x$).
- Divide the sum of all the values by the number of values (n).

Example Assume that the following numbers represent a sample of six students' test marks: 84   75   92   60   80   77

- Add the numbers $\sum x = 84 + 75 + 92 + 60 + 80 + 77 = 468$
- Divide by 6

NUMERICAL MEASURES OF DESCRIPTION

$$\bar{x} = \frac{\sum x}{n} = \frac{468}{6} = 78$$

- The arithmetic mean is 78.

Exercise 1 Determine the arithmetic mean of the following ages:

5 2 40 8 15 30 35 11

(Answer: 18.25)

Exercise 2 The estimated ages of five mummies uncovered at an excavation site were 5, 4, 9, 2 and 10 thousand years respectively. What is the mean age of this sample of five mummies?

(Answer: 6 thousand years)

Grouped data Grouped data are data that have been placed into a frequency distribution. The original data are lost, and only the frequencies within each class remain.

$$\bar{x} = \frac{\sum xf}{n}$$

Where:

x = class midpoint of each class in sample
f = frequency of each class
n = number of observations in the sample

$\sum f = n$

Steps:

- Assume that each of the items in a given class interval falls at the midpoint of that class, and compute the midpoints (x) for each class.
- Multiply each midpoint by the respective frequency of that class (xf) and sum the product: $\sum xf$.
- Sum the frequencies (f) to get the total number of observations (n).
- Divide the $\sum xf$ by n.

Example 1   Annual rainfall in Eastern Transvaal

Class interval	(1) Class midpoint (x)	(2) Frequency (f)	(3) = (1×2) xf
0–7	3.5	2	7.0
8–15	11.5	6	69.0
16–23	19.5	3	58.5
24–31	27.5	5	137.5
32–39	35.5	2	71.0
40–47	43.5	2	87.0
		$\sum f = 20$	430.0

❏ Compute midpoints (column 1) for each class by adding the lower limit to the upper limit of each class and divide by 2.

For example:

$(0 + 7) \div 2 = 3.5$; $(8 + 15) \div 2 = 11.5$; $(16 + 23) \div 2 = 19.5$

❏ Multiply midpoints by respective frequency of that class (column 1 × column 2)

$3.5 \times 2 = 7.0$; $11.5 \times 6 = 69.0$; $19.5 \times 3 = 58.5$; $27.5 \times 5 = 137.5$

❏ Add the frequency (f) (column 2), which is = n
❏ Add column 3 which is $\sum xf$

❏ Substitute into formula: $\bar{x} = \dfrac{\sum xf}{n} = \dfrac{430}{20} = 21.5$

❏ The arithmetic mean ($\bar{x}$) is 21.5

Exercise 2   The following frequency distribution was constructed from amounts spent at Rosy's by a sample of 50 customers. Determine the mean of this data.

Class interval (R)	f
0 – 9	5
10–19	12
20–29	18
30–39	8
40–49	5
50–59	2

(Answer: mean = R24.9)

NUMERICAL MEASURES OF DESCRIPTION

Note: The mean can be calculated by using a pocket calculator with a statistical programme.

Characteristics of the arithmetic mean
(1) It is commonly used and easy to calculate.
(2) Every data set has a mean. In the case of open-ended classes, intervals are assumed to have the same length as others.
(3) It is a unique value, because a set of data has only one arithmetic mean.
(4) It is reliable because it reflects all the values in the data set (but because of this, it may also be affected by extreme values that are not representative of the set as a whole).
(5) It is useful for performing statistical procedures such as hypothesis testing.

The median The median is the value that occupies the middle position of a group of numbers in a numerical order. The number of the items that lie above this point is equal to the number that lie below it.

Ungrouped data Steps:
❏ Array the data in ascending order.
❏ Determine the position of the median.

$$\text{Median position} = \frac{n+1}{2}$$

❏ If there is an odd number of items, the middle item of the array is the median. If there is an even number of items, the median is the average of the two middle items.

Example 1 Seven runners clocked the following times (in minutes) over a route run in training: 4.7, 5.0, 4.8, 5.1, 9.0, 4.3, 4.2. Determine the median.

Array: 4.2   4.3   4.7   4.8   5.0   5.1   9.0

$$\text{Median position} = \frac{n+1}{2}$$
$$= \frac{7+1}{2} = 4.0$$

Median value   = 4.8

Example 2 The numbers of patients treated in the emergency room on 8 consecutive days during the Christmas season are: 86, 52, 49, 43, 35, 31, 31, 11

$$\text{Median position} = \frac{n+1}{2} = \frac{9}{2} = 4.5$$

$$\text{Median value} = \frac{35+43}{2} = 39$$

Exercise 1 The following numbers represent the typing speeds of five secretaries in words per minute: 30, 90, 45, 25, 55. Determine the median. (Answer: 45)

Exercise 2 Compute the median of the following ages: 5, 2, 40, 8, 15, 30. (Answer: 11.5)

Grouped data With grouped data we are unable to determine where the true middle value falls, but we can estimate the median by using a formula and assuming that the median value will be the $n \div 2$ value and that the frequencies in the median class are evenly spread.

$$\text{Median} = L + \frac{(n/2 - F)\, c}{f_m}$$

Where:

$L =$ lower boundary of median class
$F =$ sum of all the class frequencies up to, but not including the median class
$f_m =$ frequency of the median class
$c =$ width of interval
$n = \sum f$

Steps:

❑ Determine the location of the median number: $n \div 2$.
❑ Construct the cumulative less-than frequency column.
❑ Compare the position of the median, that is $n \div 2$, with the cumulative frequency column to determine which one of the intervals contains the median. The median class is the interval where the cumulative frequency is equal to or exceeds $n \div 2$ for the first time.
❑ Estimate the value of the median using the formula for grouped data.

NUMERICAL MEASURES OF DESCRIPTION

Exercise 1

Class interval	f	cum. f
–0.5–7.5	2	2
7.5–15.5	6	8
15.5–23.5	3	11
23.5–31.5	5	16
31.5–39.5	2	18
39.5–47.5	2	20

- Construct the cumulative less-than frequency column (column 3).
- Determine the location of the median, the n ÷ 2 value, i.e. 20 ÷ 2 = 10th value.
- Within the first class there are 2 values. Obviously the 10th number does not fall within this class. There are 6 values in the second class, and therefore a total of 8 in the first two classes. Therefore the second class does not contain the 10th value either. Up to the end of the third class there is a total of 11 numbers. Somewhere within the third class between 15.5 and 23.5 we will find the median.

Formula: $\text{Median} = L + \dfrac{(n/2 - F)c}{fm}$

$$= 15.5 + \dfrac{(10 - 8)\,8}{3}$$

$$= 20.83$$

*Cumulative 'less-than' Ogive
to determine the Median Value*

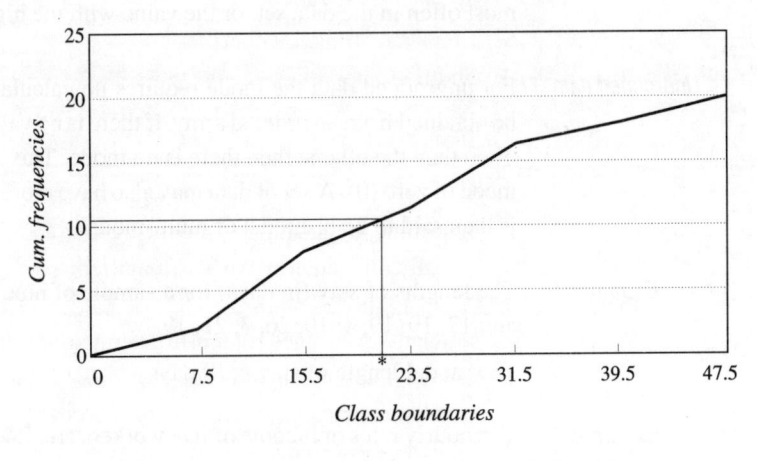

Class boundaries

— Less-than ogive   * Median value

Determine the value of the median graphically by making use of the cumulative 'less-than' ogive. The median position on the vertical axis will have a corresponding value on the horizontal axis, which is the median value.

Exercise 2   Compute the median using the following frequency distribution.

Class interval	f
5–9	1
10–14	4
15–19	8
20–24	4
25–29	3

(Answer: 17.62)

Characteristics of the median
(1) It is a better measure of central tendency than the mean when the data are very skewed.
(2) Extreme values do not affect the median as strongly as they do the mean.
(3) The median is easy to understand and can be calculated from any kind of data (quantitative or qualitative) unless the median falls in an open-ended interval.

The mode   When we talk about the 'average size' or 'average income' etc., we are not thinking of the arithmetic mean but rather of the size or income we are most likely to come across. The mode is that value which occurs most often in the data set, or the value with the highest frequency.

Ungrouped data   For ungrouped data the mode requires no calculation and can easily be obtained from an ordered array. If there is no value that occurs more often than the others, then there is no mode. This is not the same as a mode of zero (0). A set of data may also have more than one mode and is then said to be bi-modal or multi-modal.

Example 1   The lengths of stay (in days) for a sample of nine patients in ward A are: 17, 19, 19, 4, 19, 26, 3, 21, 19.

The modal length of stay is 19 days.

Example 2   The hourly rates of income of five workers are: R4, R9, R7, R16, R10.

There is no mode.

NUMERICAL MEASURES OF DESCRIPTION

Example 3   The following test scores were obtained by 8 students: 81, 39, 100, 81, 69, 76, 42, 76

The modal test scores are 76 and 81.

Grouped data   Grouped data do not show a single most frequently occuring value because the raw data are disguised in a distribution. The mode is therefore estimated using the following formula:

$$\text{Mode} = L + \left(\frac{\Delta_1}{\Delta_1 + \Delta_2}\right) c$$

Where:

L = lower boundary of modal class
Δ_1 = frequency of modal class minus f of previous class
Δ_2 = frequency of modal class minus f of following class
c = width of interval

Steps:

- Select the class containing the largest number of frequencies as the modal class.
- Use the formula to estimate the modal value.
- If the class intervals are given in class limits, remember to change the limits to boundaries first.

Example

Class interval	f
–0.5–7.5	2
7.5–15.5	6
15.5–23.5	3
23.5–31.5	5
31.5–39.5	2
39.5–47.5	2

The class with the highest frequency is the second class.

$$\text{Mode} = L + \left(\frac{\Delta_1}{\Delta_1 + \Delta_2}\right) c = 7.5 + \left(\frac{4}{4+3}\right) 8 = 12.07$$

To determine the mode graphically, the histogram of the frequency distribution is used. The modal class is the longest rectangle and the mode is located in it by:

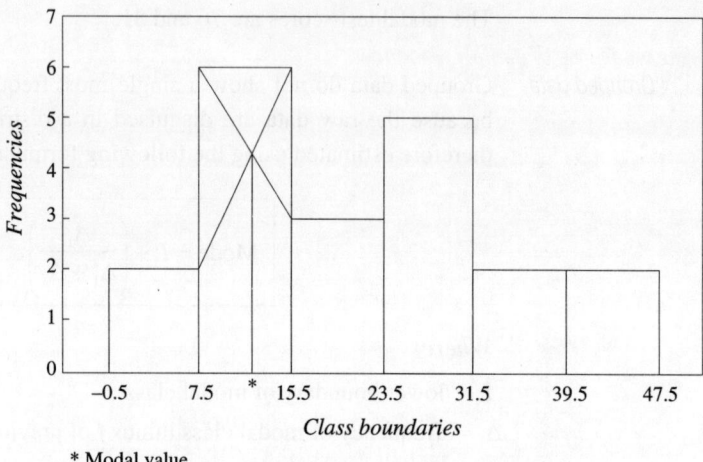

HISTOGRAM
To determine modal value

[Histogram with class boundaries at −0.5, 7.5, 15.5, 23.5, 31.5, 39.5, 47.5 and frequencies 2, 6, 3, 5, 2, 2. Modal value indicated at approximately 15.5 on the horizontal axis.]

* Modal value

- drawing a line from the top right corner of the modal rectangle to the top right corner of the rectangle to its immediate left;
- drawing a second line from the top left corner of the modal rectangle to the top left corner of the rectangle to its immediate right; and
- then drawing a straight line parallel to the vertical axis through the intersection point of the previous two lines.
- The value on the horizontal axis will approximate the modal value.

Exercise 1 Determine the modal value for the following frequency distribution:

Intervals	f
0–9	5
10–19	12
20–29	18
30–39	8
40–49	5
50–59	2

(Answer: 23.25)

Characteristics of the mode
(1) It is a very little used statistic.
(2) It does not accurately represent the data in the sample.
(3) It can be used as a central location for qualitative and quantitative data.

NUMERICAL MEASURES OF DESCRIPTION

(4) It is not affected by extreme values.
(5) It can be used together with open-ended intervals.

Other measures of central tendency

The weighted mean This measure enables us to calculate an average that takes into account the importance or weight of each value to the overall total.

$$\bar{x}_w = \frac{\sum xw}{\sum w}$$

where:
$\bar{x}_w$ = symbol for weighted mean
w = weight assigned to each observation

Example Calculate a student's final mark for statistics if we take into account 4 equally weighted semester tests which together make up 40 % of the final mark, and a 3-hour examination which makes up 60 % of the final mark. The student's results were as follows:

		w	xw
1st semester	60 %	10	600
2nd semester	47 %	10	470
3rd semester	53 %	10	530
4th semester	71 %	10	710
3-hour final	55 %	60	3 300
		100	5 610

$$\bar{x}_w = \frac{\sum xw}{\sum w} = \frac{5\,610}{100} = 56.10\,\%$$

The geometric mean This measure shows the average rate of change for values that change over a period of time.

$$GM = \sqrt[n]{x_1 \cdot x_2 \ldots x_n} = x^{1/n}$$

Example 1 Hex Textiles has shown the following percentage increase in net worth over the last five years. Determine the average percentage increase.

Year	x
1989	5
1990	10.5
1991	9
1992	6
1993	7.5

$$GM = \sqrt[5]{5 \times 10.5 \times 9 \times 6 \times 7.5} = x^{1/5}$$
$$= 21\,262.5^{1/5}$$
$$= 7.34$$

Example 2 The Light Company has manufactured the following numbers of units over the past five years. Calculate the average percentage increase over this time period in units produced,.
(The increases in units produced (x) are not given and should be determined first.)

Year	Units		x
1988	12 500		
1989	13 250	(13 250 ÷ 12 500 × 100) =	106
1990	14 310	(14 310 ÷ 13 250 × 100) =	108
1991	15 741	(15 741 ÷ 14 310 × 100) =	110
1992	17 630	(17 630 ÷ 15 741 × 100) =	112

$$GM = \sqrt[4]{106 \times 108 \times 110 \times 112}$$
$$= 141\,039\,360^{1/4}$$
$$= 108.98 \quad \text{Average increase} = 8.98\,\%$$

Harmonic mean This average is restricted to the averaging of rates where one of the variables must stay constant, e.g. in kilometers per hour, price per dozen, etc.

$$HM = \frac{n}{\sum \frac{1}{x}}$$

Example 1 If I spent R12 on eggs costing R1.20 per dozen and another R12 on eggs costing R1.50 per dozen, what was the average price per dozen?

$$HM = \frac{n}{\sum \frac{1}{x}} = \frac{2}{\frac{1}{1.5} + \frac{1}{1.2}} = 1.33$$

NUMERICAL MEASURES OF DESCRIPTION

Example 2 A car travels at 80 km/h to a certain place and back again at 60 km/h. What is the average speed per hour?

$$HM = \frac{n}{\sum \frac{1}{x}} = \frac{2}{\frac{1}{80} + \frac{1}{60}} = 68.57$$

Midrange Midrange is a value halfway between the smallest and the largest observations in a distribution.

$$\text{Midrange} = \frac{\text{largest value} + \text{smallest value}}{2}$$

4.2 MEASURES OF 'NON-CENTRAL' POSITION

These measures are used to determine the location of a specific piece or portion of the data in relation to the rest of the sample and are known as fractiles. The most familiar fractiles are quartiles, deciles and percentiles.

Fractiles that divide the data into four equal parts are called quartiles. The first quartile (Q_1) is a value such that 25 % of the observations are smaller and the third quartile (Q_3) is a value such that 75 % of the values are smaller. Deciles (D_j) divide the data into 10 equal parts and percentiles (P_j) divide the data into 100 equal parts. Because all deciles can be converted to percentiles, e.g. ($D_2 = P_{20}$) deciles will be ignored in this book.

A fractile is a value that occupies a specific position, such as the median. The exact number location must therefore be determined before the value can be calculated. The data have to be in an array.

Formulae:
Ungrouped data

$$\text{Position } Q_j = \frac{jn+2}{4}$$

$$\text{Position } P_j = \frac{jn+50}{100}$$

❏ Arrange the data into an array in ascending order.
❏ Determine the position of the fractile.
❏ Read the value of the fractile from the array.

Example The hourly wages of seven secretaries are R9.50; R3.00; R10.00; R9.50; R8.50; R7.50; R10.50.
Array: 3.00 7.50 8.50 9.50 9.50 10.00 10.50

$$Q_1 \text{ position} = \frac{1n+2}{4} = 2.25 \quad \text{therefore } Q_1 = 7.50 + 0.25(8.50 - 7.50)$$
$$= R7.75$$
$$Q_3 \text{ position} = \frac{3n+2}{4} = 5.75 \quad \text{therefore } Q_3 = 9.50 + 0.75(10 - 9.50)$$
$$= R9.88$$
$$P_{30} \text{ position} = \frac{30n+50}{100} = 2.6$$
$$P_{30} = 7.5 + 0.6\,(8.5 - 7.5) = R8.10$$

Grouped data   Assume that the frequencies in each fractile class are evenly spread.

$$Q_j = L + \frac{(jn/4 - F)c}{f_Q}$$
$$P_j = L + \frac{(jn/100 - F)c}{f_P}$$

where:
L = lower boundary of fractile class
F = sum of all the class frequencies up to but not including fractile class
f_Q and f_P = frequency of fractile class
c = width of class

Steps

- Determine the position of the quartile or percentile number, $jn/4$ or $jn/100$.
- Construct the cumulative 'less-than' frequency column.
- Compare the position of the fractile with the cumulative frequency column to determine which one of the classes contains the fractile.
- Assume that the frequencies in the fractile class are evenly spread and estimate the value of the fractile using the formula for grouped data.

Class	f	cum. f
–0.5–7.5	2	2
7.5–15.5	6	8
15.5–23.5	3	11
23.5–31.5	5	16
31.5–39.5	2	18
39.5–47.5	2	20

$$Q_1 = L + \frac{(^1n/_4 - F)c}{f_Q} = 7.5 + \frac{(5-2)8}{6} = 11.5$$

$$Q_3 = L + \frac{(^3n/_4 - F)c}{f_Q} = 23.5 + \frac{(15-11)8}{5} = 29.9$$

$$P_{40} = L + \frac{(^{40n}/_{100} - F)c}{f_P} = 7.5 + \frac{(8-2)8}{6} = 15.5$$

The values of fractiles can also be determined graphically by making use of the cumulated 'less-than' ogive.

Cumulative 'less-than' ogive to determine fractile values

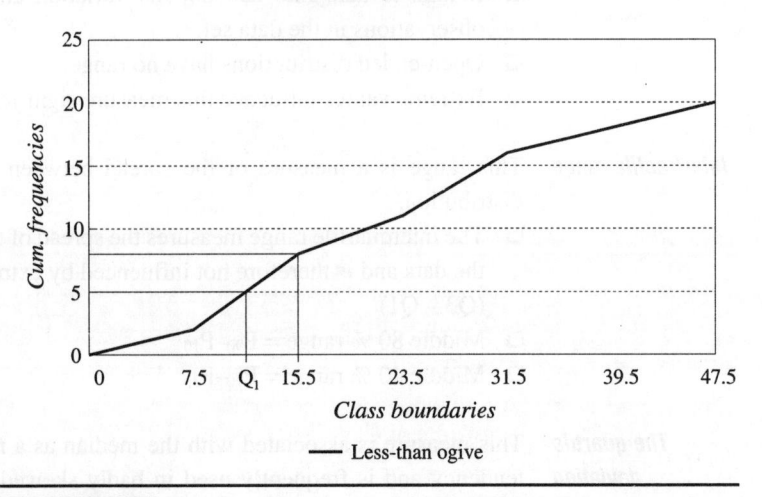

4.3 MEASURES OF DISPERSION

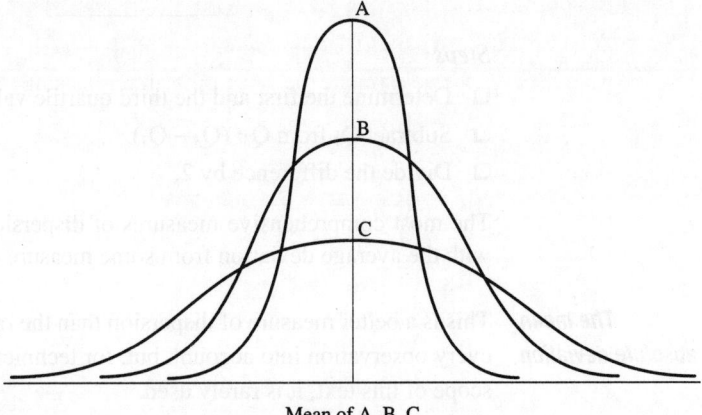

Two sets of data can have the same central location and yet be very different if one is more spread out than the other. Dispersion is an important characteristic to understand and measure because it gives additional information about the spread of the data. Curve C shows widely dispersed data. The data in curve A are more closely centred around the mean.

Different measures of dispersion

The range   The range is the difference between the highest and the lowest values in a set of data. It measures the distance across the entire set of data.

Characteristics of the range
- It is easy to understand and determine, but its usefulness as a measure of dispersion is limited.
- It fails to take into account the variation among all the other observations in the data set.
- Open-ended distributions have no range.
- Extreme values influence this measure significantly.

Interfractile range   This range is a measure of the spread between two fractiles in a distribution.
- The interquartile range measures the spread of the middle 50 % of the data and is therefore not influenced by extreme values. (Q3 – Q1)
- Middle 80 % range = P_{90}–P_{10}
- Middle 40 % range = P_{70}–P_{30}

The quartile deviation   This measure is associated with the median as a measure of central tendency and is frequently used in badly skewed distributions. Although it is a better measure of dispersion than the range, it is still a rough estimate as it ignores 50 % of the values.

$$Q_D = \frac{Q_3 - Q_1}{2}$$

Steps
- Determine the first and the third quartile values (Q_1 and Q_3).
- Subtract Q_1 from Q_3: ($Q_3 - Q_1$).
- Divide the difference by 2.

The most comprehensive measures of dispersion are those that deal with the average deviation from some measure of central tendency.

The mean absolute deviation   This is a better measure of dispersion than the ranges because it takes every observation into account, but, for technical reasons beyond the scope of this text, it is rarely used.

NUMERICAL MEASURES OF DESCRIPTION

Ungrouped data

$$AD = \frac{\sum |x - \bar{x}|}{n}$$

Steps in calculating the AD
- Calculate the arithmetic mean ($\bar{x}$) of the distribution.
- Determine the deviation of each value (x) from $\bar{x}$ without regard to the algebraic sign (i.e. absolute value): $|x - \bar{x}|$
- Add the absolute values of the deviations: $\sum |x - \bar{x}|$
- Divide the sum by the number of values (n).

Example The hourly wages (in rands) of six secretaries are:

x	$|x - \bar{x}|$
3.00	5.00
7.50	0.50
8.50	0.50
9.50	1.50
9.50	1.50
10.00	2.00
48.00	11.00

$$\bar{x} = \frac{\sum x}{n} = \frac{48.00}{6} = 8.00$$

$$AD = \frac{\sum |x - \bar{x}|}{n} = \frac{11.00}{6} = 1.83$$

Grouped data

$$AD = \frac{\sum |x - \bar{x}| f}{n}$$

Steps:
- Calculate the arithmetic mean ($\bar{x}$).
- Determine the absolute deviation of each class midpoint from the mean: $|x - \bar{x}|$
- Multiply the absolute deviation in each class by the frequency of that class: $|x - \bar{x}| f$

- Sum the absolute deviations: $\sum |x - \bar{x}| f$
- Divide the sum by the total frequencies (n)

Example

Interval	f	x	$\|x-\bar{x}\|$	$\|x-\bar{x}\| f$
0–7	2	3.5	18	36
8–15	6	11.5	10	60
16–23	3	19.5	2	6
24–31	5	27.5	6	30
32–39	2	35.5	14	28
40–47	2	43.5	22	44
	20			204

$\bar{x} = 21.5$

$$AD = \frac{\sum f|x - \bar{x}|}{n} = \frac{204}{20} = 10.2$$

Standard deviation

Characteristics
- This is the most widely used measure of dispersion about the mean.
- It includes all the numbers in the data set.
- It is based on the deviations of each x value from the mean, but it does not ignore the algebraic sign.
- It measures the square root of the average squared distances of the observations from the mean.
- The larger the standard deviation in comparison with the data size, the wider the spread of the data around the mean value.

Ungrouped data

$$s = \sqrt{\frac{\sum (x - \bar{x})^2}{n - 1}}$$

Steps
- Compute the arithmetic mean ($\bar{x}$).
- Subtract the mean from each of the individual values: $(x - \bar{x})$
- Square each difference: $(x - \bar{x})^2$
- Sum the differences: $\sum (x - \bar{x})^2$

NUMERICAL MEASURES OF DESCRIPTION

- Compute the average of this total by dividing by $n-1$. Division by $n-1$ is to correct the bias in estimating the population variance by using the sample variance.
- The standard deviation is the square root of this total.
- The units of measure are the same as the units of measure of the data.

Example The hourly wages (in rands) of six secretaries are:

x		$(x-\bar{x})^2$
3.00	$(3-8)^2 =$	25.00
7.50	$(7.5-8)^2 =$	0.25
8.50	$(8.5-8)^2 =$	0.25
9.50	$(9.5-8)^2 =$	2.25
9.50	$(9.5-8)^2 =$	2.25
10.00	$(10-8)^2 =$	4.00
		34.00

$$\bar{x} = 8.00$$

$$s = \sqrt{\frac{\Sigma(x-\bar{x})^2}{n-1}}$$

$$= \sqrt{\frac{34}{5}}$$

$$= 2.61$$

Exercise Compute the standard deviation of the following data:
84; 75; 92; 60; 80; 77
(Answer: 10.68)

Grouped data For grouped data the individual observations are not available and the total squared deviation from the mean must therefore be estimated by using the frequency occurence in each of the classes and the midpoint of each class. The estimate will be perfectly accurate if the observations within each individual class are spread symmetrically over the class.

$$s = \sqrt{\frac{\Sigma(x-\bar{x})^2 f}{n-1}}$$

STATISTICAL METHODS AND FINANCIAL CALCULATIONS

Steps:
- Compute the arithmetic mean.
- Subtract the mean from the midpoints of the individual classes and square the results.
- Multiply each squared difference by the frequency within each individual class.
- Add the results to obtain the total squared deviation from the mean.
- Compute the average of this total by dividing by $n-1$.
- The standard deviation is the square root of this total.
- The units of measure are the same as the units of measure of the data.

Example

Interval	f	x	$(x-21.5)^2$	$(x-\bar{x})^2 f$
0–7	2	3.5	324	648
8–15	6	11.5	100	600
16–23	3	19.5	4	12
24–31	5	27.5	36	180
32–39	2	35.5	196	392
40–47	2	43.5	484	968
	20			2 800

$\bar{x} = 21.5$

$$s = \sqrt{\frac{\Sigma(x-\bar{x})^2 f}{n-1}}$$

$$= \sqrt{\frac{2\,800}{20-1}}$$

$$= 12.14$$

Exercise Compute the standard deviation of the following distribution:

Interval	f
0–9	5
10–19	12
20–29	18
30–39	8
40–49	5
50–59	2

NUMERICAL MEASURES OF DESCRIPTION

(Answer: 12.61)

Note: The standard deviation can be calculated using a pocket calculator with a statistics programme.

The variance (s^2) This measure is the standard deviation squared (s^2). Although this is a very popular measure in describing data, the main drawback is that the unit of measurement is also squared.

4.4 MEASURES OF RELATIVE DISPERSION

Coefficient of variation (CV) A relative measure of dispersion is the coefficient of variation, which is the ratio, expressed as a percentage, of the standard deviation to the mean. It is used for the comparison of the variation of two sets of data with different means, sample sizes or measurement units.

The unit of measure is 'percent' and not the unit of measure of the data.

$$CV = \frac{s}{\bar{x}} \times 100$$

Example Suppose that the random variable x is the number of analyses performed by a technician per day. If the mean number of analyses per day completed by technician A is 40 with a standard deviation of 5, and the mean number of analyses per day completed by technician B is 160 with a standard deviation of 15, which employee shows less variability in the number of analyses per day?

For A: $CV = \frac{s}{\bar{x}} \times 100$

$= \frac{5}{40} \times 100$

$= 12.5 \%$

For B: $CV = \frac{s}{\bar{x}} \times 100$

$= \frac{15}{160} \times 100$

$= 9.4 \%$

So we find that B, who has more absolute variation in output than A, has less relative variation because the mean output for B is much greater than for A.

4.5 MEASURES OF SHAPE

The shape of the frequency distribution can be described by:
(1) its symmetry or lack of it (skewness), and
(2) its peakedness (kurtosis)

Skewness (SK) Skewness is a frequency distribution's degree of distortion from symmetry. The skewness of a symmetrical distribution is zero because the frequency density tapers off equally in both directions from the mode. Positive skewness is so called because the frequency density tapers off more slowly toward the right of the mode than toward the left, and the long right-hand tail of the frequency curve points in the positive direction along the horizontal axis. Negative skewness is so called because the frequency density tapers off more slowly toward the left of the mode than toward the right, and the long left-hand tail points in the negative direction along the horizontal axis. The shape of a distribution can be determined graphically using the histogram or polygon.

The degree of skewness should be calculated to determine if a frequency distribution can be described by its mean and standard deviation, as these two measures are most useful when describing a symmetrical or normal distribution. A large degree of skewness indicates that the distribution does not approach a normal distribution.

Interestingly, the type of skewness has certain implications for the positions of the mean, median and mode.

(1) In the case of zero skewness there is a symmetrical distribution, where the mean = median = mode.

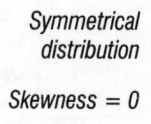

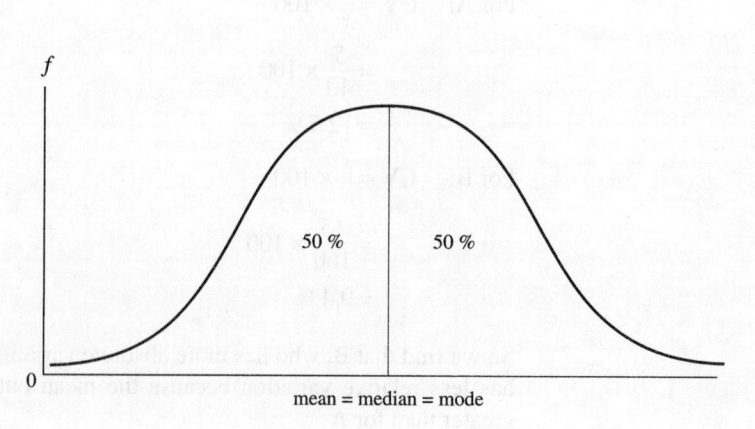

Symmetrical distribution

Skewness = 0

50 %   50 %

mean = median = mode

NUMERICAL MEASURES OF DESCRIPTION

(2) Positive skewness occurs when the mean is increased by the addition of some extremely large values to the data set, causing the mean to be pulled towards these extreme values.

The median, being dependent on the number of items in the data set rather than on the size of those items, is less sensitive than the mean, and ends up somewhere between the mode and the mean. The mode is not influenced by extreme values at all and stays at the peak of the distribution. The median is typically closer to the mean than to the mode: for moderately skewed distributions, the median lies about one third of the way between the mean and mode.

Positively skewed distribution

Skewness > 0

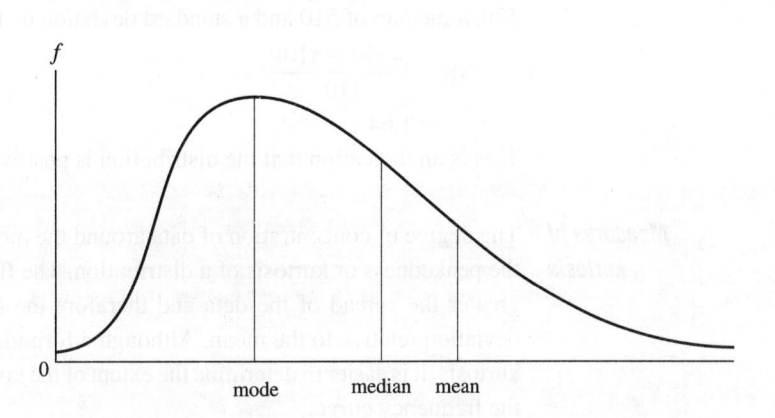

(3) The converse is true for negatively skewed distributions.

Negatively skewed distribution

Skewness < 0

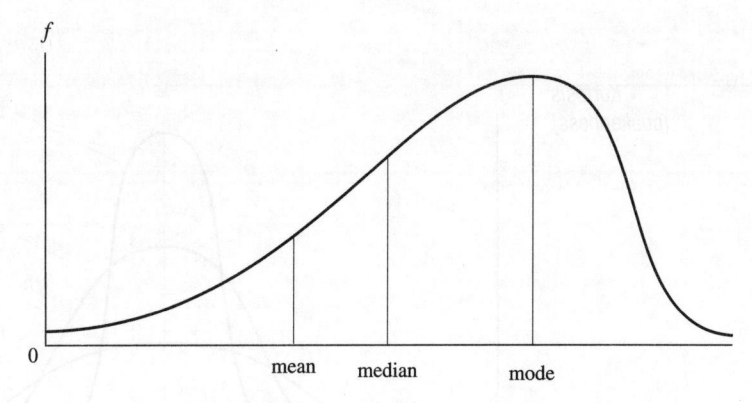

The positional differences among mean, median and mode can be used to create arithmetic measures of skewness.

Pearson's first coefficient of skewness

$$SK = \frac{\text{mean} - \text{mode}}{\text{standard deviation}}$$

Pearson's second coefficient of skewness

$$SK = \frac{3(\text{mean} - \text{median})}{\text{standard deviation}}$$

If mean = median = mode, SK = 0
If mean > median > mode, SK > 0 but smaller than 3.
If mean < median < mode, SK < 0 but greater than −3.

Example Calculate Pearson's second coefficient of skewness using a mean of 570, a median of 510 and a standard deviation of 110.

$$SK = \frac{3(570 - 510)}{110}$$

$$= 1.64$$

This is an indication that the distribution is positively skewed.

Measures of kurtosis The degree of concentration of data around the mode is referred to as the peakedness or kurtosis of a distribution. The flatter the curve, the greater the spread of the data and therefore the larger the standard deviation relative to the mean. Although a formula exists to measure kurtosis, it is easier to determine the extent of the kurtosis by observing the frequency curve.

The mesokurtic curve indicates a normal distribution.
The platykurtic curve indicates a large dispersion.
The leptokurtic curve indicates a small dispersion.

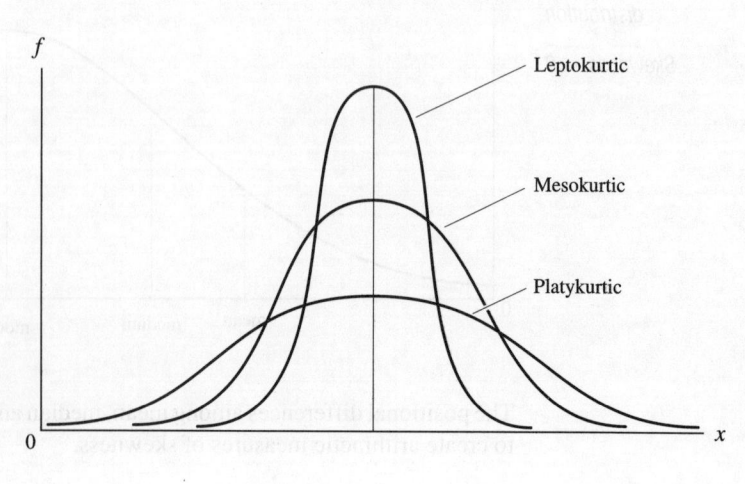

Kurtosis (peakedness)

EXERCISES

Exercise 1 The following are the weekly amounts (in rands) of welfare payments made to a sample of seven families, each with six children or more.

Family:	Asel	Kraft	Smith	Klein	Dorse	Bell	Steyn
Amount:	139	136	140	130	147	136	138

Determine the mean, median and modal weekly payments.

Exercise 2 The starting annual salaries (in thousands of rands) for a sample of six recent graduates are shown below. Compute the mean, median and modal starting salaries.

16.5   14.7   19.0   18.5   20.0   19.3

Exercise 3 Determine the median of the following set of data:

1.08   0.98   0.97   1.10   1.03   1.13   1.07   1.24   0.99   1.13
0.99   1.43   1.18   1.02   1.12   1.17   0.98   1.28   0.98   1.09

Exercise 4 For the frequency distribution below, determine the mean, median and mode.

Class interval	Frequency
100–149.5	12
150–199.5	14
200–249.5	27
250–299.5	58
300–349.5	72
350–399.5	63
400–449.5	36
450–499.5	18

Exercise 5 For the following data, calculate the mean, median and mode.

Class	f
0–24.9	6
25–49.9	11
50–74.9	14
75–99.9	16

Exercise 6 Determine the mode of the following sample:

5 8 11 9 8 6 8 7 12 8
7 7 11 8 6 10 13 7 8

Exercise 7 The gross displacement of each ship utilizing the Panama Canal during a one week period was compiled into the frequency distribution below.

Gross Displacement ('000 tons)	Frequency
0–2.99	30
3–5.99	47
6–8.99	69
9–11.99	32
12–14.99	18
15–17.99	3
18–20.99	1

Compute the sample mean, median and mode of this data.

Exercise 8 A survey of 20 households on the quality of a particular TV programme yielded the following distribution of ratings (positive numbers indicating a favourable rating and negative numbers indicating an unfavourable one):

Quality rating	Frequency
−30.0 to −20.1	3
−20.0 to −10.1	7
−10.0 to −0.1	5
0.0 to 9.9	2
10.0 to 19.9	2
20.0 to 29.9	1

Determine the mean, median and modal ratings given to the TV programme.

Exercise 9 In a rare discovery in Egypt, 16 small statues were found, presumed to be the first from a much larger population of similar statues yet to be discovered. The heights (in centimetres) of the statues are:
12.1 12.0 11.7 12.4 12.1 12.2 12.0 12.3
11.9 11.8 12.3 11.9 12.2 12.1 11.8 12.0
Calculate their mean, median and modal height.

Exercise 10 A study of the number of fares on a particular day for a sample of 40 taxi drivers revealed the following:

Number of fares	Frequency
0–4	3
5–9	6

NUMERICAL MEASURES OF DESCRIPTION

Number of fares	Frequency
10–14	8
15–19	13
20–24	7
25–29	3

By making use of the appropriate graphs, estimate the median and modal number of fares.

Exercise 11 If a person invests R2 000 at 7 %, R5 000 at 8 % and R25 000 at 9 %, what is the average return on the total investment?

Exercise 12 A lecturer gives four tests during a semester, and in calculating the final mark, weights them in proportion to their length. If a student scores 66 on a one-hour test, 72 on another one-hour test, 85 on a 20-minute test and 50 on the two-hour final, what would be his mean mark on the four tests?

Exercise 13 In one year a large employer made plan A awards averaging R4 200 to 150 employees and plan B awards averaging R4 900 to 350 employees. What was the average award made to these employees?

Exercise 14 If some objects with a mean mass of 20 kg have a total mass of 200 kg, and some other objects with a mean mass of 35 kg have a total mass of 700 kg, what is the mean mass of all the objects combined?

Exercise 15 If a man spends R12 on novelty items costing 40 c per dozen and another R12 on other items costing 60 c per dozen, what did he spend on average per dozen items purchased?

Exercise 16 If a baker buys R72 worth of honey at 60 c per bottle, R72 worth at 72 c per bottle, and R72 worth at 90 c per bottle, what was the average cost per bottle bought?

Exercise 17 An electronics firm has two divisions, one having 25 000 employees, and the other 5 000. In the first the employees averaged 42.15 hours per week, in the second 36.25 hours per week. Calculate the averaged hours worked per week for the entire firm.

Exercise 18 In a company having 80 employees, 60 earned R3.00 per hour and 20 earned R2.00 per hour. Determine the mean earnings per hour.

Exercise 19 The bacterial count in a certain culture increased from 1 000 to 4 000 in three days. What was the average percentage increase per day?

Exercise 20 A man travels from A to B at an average speed of 30 km/h and returned from B to A along the same route at an average speed of 60 km/h. Determine the average speed for the entire trip.

Exercise 21 If the price of a commodity doubles over a period of four years, what is the average percentage increase per year?

Exercise 22 In 1950 and 1960 the population of a state was 151.3 and 170.3 million respectively. What was the average percentage increase per year?

Exercise 23 Cities A, B and C are equidistant from each other. A motorist travels from A to B at 30 km/h, from B to C at 40 km/h and from C to A at 50 km/h. Determine his average speed for the entire trip.

Exercise 24 At harvest time a farmer employed 20 women, 10 men and 16 boys to lift potatoes. The women's work was three-quarters as effective as that of the men, while the boys' work was only half as effective as that of the men. Calculate the daily wage bill if a man's rate was R2.40 per day and the rate for women and boys in proportion to their effectiveness. Calculate the average daily rate for the 46 workers.

Exercise 25 Net profit after tax of Metric Systems for the past five years were R20 000, R50 000, R60 000, R80 000 and R120 000. The chairman wants to determine the average annual percentage increase in profits for inclusion in a report to the stockholders at the next annual meeting. What should he include in his report?

Exercise 26 The total number of industrial accidents occurring in the production division of Continental Motors over the past five years is given below:

Year:	1990	1991	1992	1993	1994
Number:	300	315	375	420	460

Determine the average yearly percentage increase in accidents over the period.

Exercise 27 If the number of business failures have been determined for the five years from 1989 to 1993 as follows, determine the average percentage decrease.

1989	1990	1991	1992	1993
9000	8900	8400	8300	8100

Exercise 28 Barnes Textiles has shown the following percentage increase in net worth over the last five years:

NUMERICAL MEASURES OF DESCRIPTION

1990	1991	1992	1993	1994
5 %	10.5 %	9 %	6 %	7.5 %

What was the average percentage increase in net worth over the five year period?

Exercise 29   Over a three-year period, a store owner purchased R60 worth of acrylic sheeting for new display cases in three equal purchases of R20 each. The first purchase was at R1 per metre, the second R1.10 and the third R1.15. What was the average price per metre paid for all the sheeting?

Exercise 30   A radar check of five cars travelling on the highway recorded these speeds in kilometres per hour:
86   70   91   110   89
The speeds of five other cars recorded at about the same time on a nearby highway were:
89   87   89   85   92
Compute the mean and the range of each set of data and compare the dispersion for the two sets of speed checks.

Exercise 31   The test kitchen of Cellulite, a large producer of cake mixes, constantly monitors the mass, moisture content and flavour of its cakes. The masses (in grams) of five peach upside-down cakes are 498, 500, 503, 498 and 501 respectively
(a) What is the absolute mean deviation and how is it interpreted?
(b) What are the sample variance and the standard deviation?

Exercise 32   The ages for a sample of ten women who started in new jobs are 18, 25, 31, 19, 22, 21, 19, 25, 16 and 27 respectively. Compute the following descriptive measures.
(a) the range;
(b) the absolute average deviation;
(c) the standard deviation;
(d) the coefficient of variation;
(e) the coefficient of skewness;
(f) the interquartile range; and
(g) the quartile deviation.

Exercise 33   The ground gained in rucks by the Rams rugby team during the first seven games of the season was 210, 203, 162, 134, 390, 184 and 211 metres respectively. Compute:
(a) the middle 80 % range;
(b) the interquartile range;
(c) the absolute mean deviation;

(d) the coefficient of variation; and
(e) Pearson's first and second measures of skewness.

Exercise 34 South Africa was divided into 25 statistical regions, and the percentage of households in which a female was the head of the household was determined for each region:

Percentage of households	Frequency
5–9	5
10–14	7
15–19	9
20–24	3
25–29	1

Determine the following descriptive measures:

(a) the range and midrange;
(b) the middle 90 % range;
(c) the standard deviation and variance;
(d) the coefficient of variation;
(e) the quartile deviation;
(f) the interquartile range; and
(g) the first and second coefficients of skewness of Pearson, together with an appropriate sketch to interpret the results.

Exercise 35 The following frequency distribution was compiled from the records of Fedline, a local firm providing pet care.

Number of customers per day	Frequency
0–5	6
6–11	10
12–17	25
18–23	35
24–29	20
30–35	13

Compute:

(a) the coefficient of variation;
(b) the quartile deviation; and
(c) the skewness. Interpret your result.

Exercise 36 At a certain university junior lecturers have a mean salary of R45 000 with a standard deviation of R15 000, while professors have a mean salary of R120 000 with a standard deviation of R30 000. Which group has the greater relative dispersion?

Exercise 37 Two growers of grapefruit have obtained the following statistics regarding the mass of their current crops
Grower M: $\bar{x} = 300$ g with s = 20 g
Grower P: $\bar{x} = 280$ g with s = 40 g
(a) Which grower's grapefruit are more uniform in mass?
(b) Which grower's grapefruit are probably larger?

Exercise 38 The financial controller for the Bacchus Wine Company has the company's short-term cash in a variety of savings accounts with the following interest rates:
5.25 %   5.5 %   5.75 %   6 %   6.5 %   7 %
Calculate the mean, variance and standard deviation of these rates.

Exercise 39 The Martin Rubber Company has two factories, A and B. Both factories employ many high school students in the summer. In A, the students average R196.40 per week with a standard deviation of R31.60. In B, the students average R241.60 per week with a standard deviation of R42.80. Which plant has the greatest relative dispersion?

Exercise 40 The administrator of a hospital conducted a survey of the number of days patients stayed in the hospital following an operation.

Days in hospital	Frequency
1–3	32
4–6	108
7–9	67
10–12	28
13–15	14
16–18	7
19–21	3
22–24	1

(a) Calculate the mean and standard deviation.
(b) How many patients stay less than 12 days?
(c) How many patients stay more than 17 days? Make use of an appropriate graph to estimate your answers for both (b) and (c).
(d) Estimate the modal number of days by making use of a graph.
(e) Discuss the shape of the distribution.
(f) Determine the quartile deviation.

CHAPTER 5
An Introduction to Probability

CONTENTS

    5.1 CONCEPTS IN PROBABILITY 75
    5.2 ELEMENTARY PROBABILITY RULES 76
    5.3 PRINCIPLES OF COUNTING 80

Personal and managerial decisions are made in the face of uncertainty. Our need to cope with uncertainty leads us to the study and use of probability theory.

Probability is the yardstick used to measure the reasonableness of expecting a particular sample result to be valid for the whole population. It is expressed as a numerical measure (on a scale between 0 and 1) of the likelihood of occurrence of an uncertain event. Decisions can then be based on this measure of the probability that a particular event will occur.

$$\text{Probability} = \frac{\text{Total number of successes}}{\text{Total number of outcomes}}$$

The process of reasoning from a set of sample observations to a general conclusion about a population is known as statistical inference.

5.1 CONCEPTS IN PROBABILITY

Experiment: the act or observation of taking some type of measurement.
Outcome: a particular result of an experiment.
Sample space: the set of all possible outcomes of the experiment.
Event: any set of outcomes of an experiment.

Example *Experiment:* One roll of a single, six-sided die.
Sample space: S = (1, 2, 3, 4, 5, 6)
Outcome: May be a 1 or a 2 or a 3 etc.
Event of odd numbers: A = (1, 3, 5)

5.2 ELEMENTARY PROBABILITY RULES

(1) Probability is measured according to a scale marked zero at one end and one at the other. Therefore a probability cannot be negative and nor can it exceed 1.
If P(A) = 0, then event A has no chance of occurring.
If P(A) = 1, then event A is certain to occur.

(2) If the probability of the occurrence of event A is P(A) and the probability that event A will not occur is P($\overline{A}$), then:

$$P(A) = 1 - (PA)$$
or
$$(PA) = 1 - P(A)$$

Example Number of children: 0   1   2   3   4   5   6 or more
Proportion of families having this many children:
0.05   0.10   0.30   0.25   0.15   0.10   0.05

The probability of a family having 5 or fewer children is most easily obtained by subtracting from 1 the probability of the family having 6 or more children, and this is seen to be:

1 − 0.05 = 0.95

In the study of probability it is often necessary to combine the probabilities of events by making use of addition or multiplication rules. Addition rules involve the addition of probabilities, and multiplication rules involve the multiplication of probabilities

Special rule of addition The special rule of addition states that the probability of the event A or the event B occurring in a given observation is equal to the probability of event A plus the probability of event B.

$$P(A \text{ or } B) = P(A) + P(B)$$

To apply the special rule of addition, the events must be mutually exclusive. This means that when one of the events occurs, none of the others can occur at the same time.

AN INTRODUCTION TO PROBABILITY

Example 1 If there are 10 horses in a race, all with an equal chance of winning, what is the probability that horse number 3 or 5 or 8 will win?
Note: The formula can be extended.

$$P(3 \text{ or } 5 \text{ or } 8) = \frac{1}{10} + \frac{1}{10} + \frac{1}{10}$$

$$= \frac{3}{10}$$

Example 2 A securities analyst maintains that there is a 0.3 chance that a certain company will be bought up by Exxon and a 0.3 chance that it will be purchased by Mobil. What is the probability that it will be acquired by one or the other of these two companies?

$$P(\text{Exxon or Mobil}) = 0.3 + 0.3$$
$$= 0.6$$

General rule of addition If two events are not mutually exclusive, both events may occur simultaneously in a single observation. To avoid double counting, the probability of the occurrence of an A or a B is reduced by the probability of the occurrence of both of them together.

$$\boxed{P(A \text{ or } B) = P(A) + P(B) - P(A \text{ and } B)}$$

Example 1 In determining the probability of drawing an ace or a spade from a pack of 52 cards, the possibility of drawing the ace of spades (i.e. an ace and a spade) must be taken into account.

$$P(\text{ace or spade}) = \frac{4}{52} + \frac{13}{52} - \frac{1}{52}$$

$$= \frac{16}{52}$$

Example 2 What is the probability, on one roll of a pair of dice, of obtaining a sum of ten or a double (i.e. the same number showing on both dice)?

$$P(10 \text{ or double}) = \frac{3}{36} + \frac{6}{36} - \frac{1}{36}$$

$$= \frac{8}{36}$$

Special rule of multiplication The special rule of multiplication is used to find the probability of the simultaneous or successive occurrence of two (or more) independent or unrelated events A and B, that is, events where the occurrence of event A does not affect the occurrence of event B.

$$P(A \text{ and } B) = P(A) \times P(B)$$

Example 1 An urn contains 14 red marbles and 6 green marbles. If sampling with replacement is done, what is the probability that a sample of two will contain two red marbles?

$$P(\text{red and red}) = \frac{14}{20} \times \frac{14}{20}$$
$$= 0.49$$

Example 2 The history of a manufacturing process shows that on average 5 out of every 100 items produced have been defective. If two items are randomly selected with replacement, what is the probability that both will be defective?

$$P(A \text{ and } B) = 0.05 \times 0.05$$
$$= 0.0025$$

General rule of multiplication This rule is used to combine events that are dependent on each other. For two events, the probability of the second event is affected by the outcome of the first. The probability of both A and B occurring is:

$$P(A \text{ and } B) = P(A) \times P(B|A)$$
or
$$P(A \text{ and } B) = P(B) \times P(A|B)$$

where $P(P|A)$ is the conditional probability. A conditional probability is the likelihood of a second event occurring, given that the first has already occurred.

Example If a box contains 6 red marbles and 4 blue marbles, what is the probability of drawing a blue marble in a second draw if the first marble drawn was blue? What is the probability of drawing 2 blue marbles

$$P(B_1) = \frac{4}{10}$$
$$P(B_2|B_1) = \frac{3}{9}$$
$$P(B_1 \text{ and } B_2) = \frac{4}{10} \times \frac{3}{9}$$
$$= \frac{12}{90}$$

AN INTRODUCTION TO PROBABILITY

Example Whenever a new home is purchased, Dirk Insurance Company mails an advertisement for bond insurance to the new homeowner. The company encloses a coupon for the homeowner to return if more information is desired. If the coupon is returned, a sales person will call on the customer. Records show that 1 out of 100 advertisements results in a response and that a salesperson can sell insurance to 1 out of every 10 customers visited. What is the probability that a new homeowner will purchase bond insurance from Dirk Insurance Company?

$$P(\text{return and purchase}) = P(\text{return}) \times P(\text{purchase|return})$$
$$= \frac{1}{100} \times \frac{1}{10}$$
$$= 0.001$$

Tree diagram A tree diagram is a schematic representation of the various possible outcomes of an experiment. Each column in the diagram shows the number of trials, and the line segments from the starting point give the number of possible combinations together with their probabilities. The probabilities should total 1.

Example A jar contains 4 red balls and 6 green balls. If two balls are drawn from the jar without replacement, list all the possible outcomes together with their probabilities

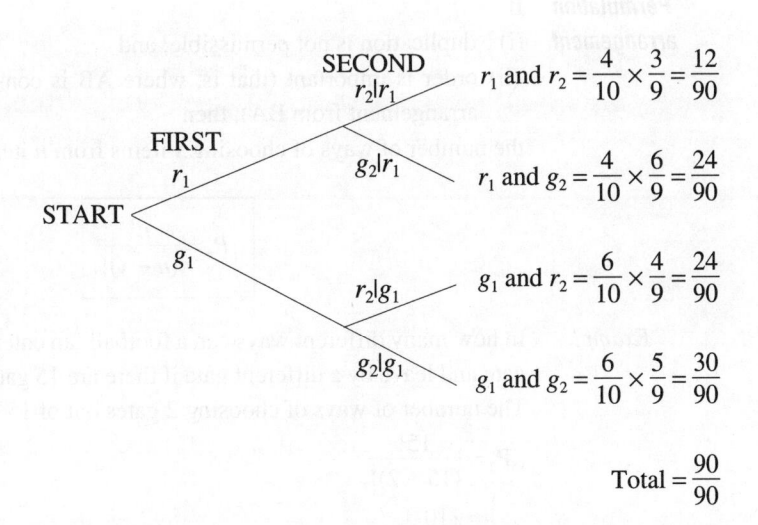

```
                          SECOND
                          r₂|r₁        r₁ and r₂ = 4/10 × 3/9 = 12/90
            FIRST
             r₁           g₂|r₁        r₁ and g₂ = 4/10 × 6/9 = 24/90
   START
             g₁
                          r₂|g₁        g₁ and r₂ = 6/10 × 4/9 = 24/90

                          g₂|g₁        g₁ and g₂ = 6/10 × 5/9 = 30/90

                                                    Total = 90/90
```

5.3 PRINCIPLES OF COUNTING

Probability is based on the number of possible outcomes that make up the numerator and denominator.

$$\text{Probability} = \frac{\text{number of successes}}{\text{total number of outcomes}}$$

The following methods are applied in counting the possible outcomes:

- the multiplication method;
- the permutation arrangement; and
- the combination arrangement.

Multiplication method If a particular experiment has n_1 possible outcomes on the first trial and n_2 possible outcomes on the second, and duplication is permissible, the total number of outcomes for the two trials are

$$n_1 \times n_2 \times \ldots n_j$$

Example Suppose a car registration number consists of two letters followed by three numbers of which the first number is not zero. How many different car number plates can be printed?

L L N N N

$26 \times 26 \times 9 \times 10 \times 10 = 608\,400$

Permutation arrangement If
(i) duplication is not permissible; and
(ii) order is important (that is, where AB is considered a different arrangement from BA); then

the number of ways of choosing x items from n items is:

$$_nP_x = \frac{n!}{(n-x)!}$$

Example In how many different ways can a football fan enter a stadium by one gate and leave by a different gate if there are 15 gates in the stadium? The number of ways of choosing 2 gates out of 15 gates is:

$$_{15}P_2 = \frac{15!}{(15-2)!}$$

$$= 210$$

AN INTRODUCTION TO PROBABILITY

Combination arrangement If
(i) duplication is not permissible; and
(ii) order is not important (that is, where AB is considered the same arrangement as BA); then

the number of possible arrangements from n taken x at a time is:

$$_nC_x = \frac{n!}{x!(n-x)!}$$

Example How many different combinations of 2 cards can be made with a total of 5 cards?

$$_5C_2 = \frac{5!}{2!(5-2)!}$$
$$= 10$$

EXERCISES

Special rule of addition $P(A \text{ or } B) = P(A) + P(B)$

Exercise 1 Five students went for job interviews with a company that has announced that it will hire only one of five by random drawing. The group consists of Brian, Heila, John, Sue and Wally. What is the probability that either John or Sue will be hired?

Exercise 2 What is the probability of drawing a red or a green ball from a jar containing three red, four green and five blue balls?
What is the probability of drawing a red, green or a blue ball?

Exercise 3 What is the probability that an even number will result from one roll of a die?

Exercise 4 Two hundred randomly selected prisoners from a high-security prison are surveyed and classified by type of crime committed. Assume the type of crime to be mutually exclusive.

Murder	Robbery	Rape	Kidnapping	Other
48	42	101	7	2

What is the probability that a particular prisoner selected from the sample is:

(a) a murderer,

(b) a murderer or a kidnapper?

Exercise 5 A study was made of 138 children brought to hospitals after having been abused by adults. The following table shows the abused children's positions in their respective families:

Position	Frequency
Only	34
Oldest	24
Youngest	50
Other	30

What is the probability that a randomly selected child from this group will be:

(a) either the oldest or the youngest in the family?

(b) not the only child?

General rule of addition   P(A or B) = P(A) + P(B) − P(A and B)

Exercise 6 A welfare worker is studying the residents of a certain retirement community. She finds that 20 % of the residents receive disability payments, and 85 % receive retirement incomes. Fifteen per cent receive both disability and retirement incomes. If a resident is randomly chosen, what is the probability that that person receives disability payment or retirement income (or possibly both)?

Exercise 7 An analysis of student records at a university revealed that 45 % of the students have average marks of more than 60 %. Twenty five per cent of the students have jobs. Ten percent of the students have jobs and have average marks of more than 60 %. What is the probability that a student selected at random will have an average mark of more than 60 % or have a job?

Exercise 8 What is the probability of drawing either an ace or a heart from a pack of 52 playing cards?

AN INTRODUCTION TO PROBABILITY

Exercise 9   Twenty per cent of directors on boards of large companies are women. Five per cent are persons connected with universities. Two per cent of large companies have female board members with university ties. What is the probability that the board of directors of a randomly selected company will have a member who is a woman or someone from a university?

Exercise 10   A company established that 40 % of its customers read *Newsweek*, 30 % read *Time*, and 55 % read at least one of the two news magazines. What is the probability that a randomly selected customer will read both magazines?

Special rule of multiplication $P(A \text{ and } B) = P(A) \times P(B)$

Exercise 11   An investment service recommends the purchase of one stock and one bond independent from one another. Both will increase in price over the next year. If 70 % of all stocks and 60 % of all bonds increased in price, what is the probability that by mere random selection, the investment service would have picked both a stock and a bond that increased in price?

Exercise 12   Suppose a red and a white die are rolled. What is the probability of obtaining a pair of twos?

Exercise 13   Out of 100 cars that start in the Grand Prix race, only 60 finish. Two cars are entered by the Total team in this year's race. What is the probability that:
(a) both cars will finish?
(b) neither of the two will finish?

Exercise 14   The Post Bank has two computers. The probability that the newer one will break down in any particular month is 0.05. The probability that the older one will break down is 0.10. What is the probability that both will break down during July?

Exercise 15   Three people are working independently at solving a statistics problem. The probability of each solving the problem is ¼, ½ and ⅓ respectively. What is the probability that the problem will be solved?

General rule of multiplication $P(A \text{ and } B) = P(A) \times P(B|A)$ or $P(A \text{ and } B) = P(B) \times P(A|B)$

Exercise 16 The Bunte Tune-up Centre has received a shipment of four carburettors. One is known to be defective. If two are selected at random and tested, what is the probability that:

(a) neither is defective?
(b) the defective carburettor is located?

Exercise 17 Ten students are being interviewed for appointment to the students council. Six of them are female and four are male. Two are selected at random to be interviewed tomorrow. What is the probability that:

(a) at least one is female?
(b) at least one is male?

Exercise 18 A certain product has two batteries. The probability that the first battery will run down is 0.30. The probability that both batteries will run down is 0.06. If the first battery is found to be flat, what is the probability that the second battery will be flat?

Exercise 19 A student is interviewed for a job at Karco. The probability that, after the interview, he will want the job is 0.88. The probability that Karco will want him is 0.45. The probability that he will want the job if Karco wants him is 0.92. What is the probability that the student will want the job and that Karco will want him? What is the probability that Karco will want the student if the student wants the job?

Exercise 20 A sales promotion letter is sent to prospective clients stating that if they return it they will receive a free gift and a sales demonstration. If past records show that 10 % of customers return the letter, and 25 % of customers who see the sales demonstration purchase the product, what is the probability that a letter will result in a sale?

Counting rules

Multiplication rule $M(n, x) = n_1 \cdot n_2 \cdot n_3 \cdot n_4 \ldots n_x$

Exercise 21 If a licence number consists of three digits followed by two letters, what is the total number of possible licence numbers?

Exercise 22 If a restaurant menu had a choice of four appetizers, ten entrées, three beverages and six desserts, what is the total number of possible dinners?

Exercise 23 If a coin is tossed seven times, how many different outcomes are possible?

AN INTRODUCTION TO PROBABILITY

Exercise 24 The Weekly Double at the local race track consists of picking the winners of the first two races. If there are ten horses in the first race and thirteen in the second, how many Weekly Double possibilities are there?

Exercise 25 A gardener has seven rows available in his vegetable garden to plant tomatoes, eggplant, cucumbers, beans, peppers, lettuce and squash. Each vegetable will be allotted one row only. In how many ways can these vegetables be positioned in the garden?

Permutations $$P(n,x) = \frac{n!}{(n-x)!}$$

Exercise 26 The Big Triple at the local race track consists of picking, in the correct order, the first three horses in the ninth race. How many possible Big Triple outcomes are there if the ninth race is run by 12 horses?

Exercise 27 A rugby team must schedule a game with each of three other teams. There are five dates available for games. How many different schedules can be arranged?

Exercise 28 Mr Kuun, a builder, has eight basic house designs and five plots on Mall Road. Zoning regulations in his community do not permit look-alike houses on the same street. In how many different ways can the new houses be arranged?

Exercise 29 Three bursaries are available to students at a certain college. The values are R5 000, R6 000 and R7 000. If 15 qualified students applied, in how many different ways could these bursaries be awarded?

Exercise 30 A tourist in Egypt wants to visit four of sixteen archaeological sites. If the order of the visits matters, in how many ways can this person plan the trip?

Combinations $$C(n,x) = \frac{n!}{x!(n-x)!}$$

Exercise 31 The Drummer Club wants to select two people to serve as a membership committee. How many different committees could be selected if there are five candidates?

Exercise 32 A group of seven mountain climbers wishes to form a mountain climbing team of five. How many different teams could be formed?

Exercise 33 The director of welfare has just received 20 new welfare cases for investigation. A social worker will be assigned eight new cases. How many different groups of eight cases could be formed?

Exercise 34 How many different poker hands are possible from a pack of 52 cards? (A hand consists of five cards.)

Exercise 35 A mail-order company sells eight different books. As part of a special promotion, customers may select three different books to make up a package. How many different packages are possible?

General exercises

Exercise 36 A city council is made up of eight members of whom two are teachers. If two councillors are selected at random to fill vacancies on the training committee, what is the probability that both the teachers will be selected?

Exercise 37 Two cards are drawn from a pack of 52. Calculate the probability that a draw will include an ace and a ten?

Exercise 38 You are not aware of it, but in a case of wine you bought, five of the twelve bottles are bad. If you were to select two bottles from the case, what is the probability that:
(a) both bottles are bad?
(b) both bottles are good?
(c) the first bottle selected is bad and the second good?
(d) the first bottle selected is good and the second bad?
(e) one of the two bottles is bad?

Exercise 39 A smoke detector system uses two devices, A and B. If smoke is present, the probability that it will be detected by device A is 0.95, by device B 0.98 and by both devices 0.94.
(a) If smoke is present, what is the probability that the smoke will be detected by either A or B or both devices?
(b) What is the probability that the smoke will go undetected?

Exercise 40 One radio in a shipment of four is defective. If a dealer selects two radios at random to display in his store, what is the probability that exactly one of the radios is defective?

Exercise 41 A taste-testing experiment is conducted in a local grocery store. Two brands of margarine are tasted by a passing shopper who is then asked to state a preference for brand C or brand P. Suppose that four shoppers are asked to participate in the experiment, what is the probability that all four shoppers will choose brand P?

Exercise 42 A dealer who buys items in lots of ten, selects two of the items at random and inspects them thoroughly. He accepts all ten if there are no defects in the two inspected. Suppose that a lot contains two defective items, what is the probability that the dealer will:
(*a*) accept all ten?
(*b*) find both defective items?

Exercise 43 An experiment involves ranking three applicants in order of merit. In how many ways can the applicants be ranked?

Exercise 44 Four items in a lot are good and two are defective.
(*a*) How many different samples can be formed by selecting two items from these six?
(*b*) How many samples will consist of one good and one defective item?
(*c*) What is the probability that one good and one defective item can be drawn?

Exercise 45 In how many ways can three different office positions be filled if there are seven applicants who are qualified for all three positions?

Exercise 46 A company selects two sites from 10 available sites under consideration for the building of two new factories. If one factory will produce flash bulbs and the other cameras, in how many ways can the selection be made?

Exercise 47 Thirteen company employees have been found to be equally qualified for promotion to a particular job. It has been decided to choose five of the employees at random for promotion. How many different groups of five are possible?

Exercise 48 There are six new advertising accounts that the manager of an advertising agency must assign to his six employees. In how many different ways can the six accounts be assigned?

Exercise 49 A company makes six different models of radios. A magazine advertisement is being prepared and the layout provides space for displaying only four of the radios. It has already been decided that the most expensive and a medium-priced model will definitely appear in the advertisement. If the other radios are selected at random, how many different layouts are possible?

Exercise 50 Find the number of ways in which a tennis team consisting of four people can be chosen from a group of ten people?

Exercise 51 A salesman sells two products, A and B. He makes four calls per day. If the chance of making a sale of product A at each call is one in five and the corresponding chance of making a sale of product B at each call is one in four, calculate the probability that he will sell product A and product B at the first call.

Exercise 52 An agency has six advertisements of which any three must be placed in the April issue of a magazine. In how many different ways can the advertisements be placed if the order in which they are placed is of importance?

Exercise 53 Genco has raised the price of calculators in four out of the past six years. In three out of the four times that Genco has raised its price, its chief competitor, Texo, has raised its price too. Texo has never raised its price unless Genco increased its price first. What is the probability:

(a) that Genco will raise its price next year;

(b) Genco and Texo will raise their prices; or that

(c) Genco or Texo will raise its price; or that

(d) neither Genco nor Texo will raise its price?

Exercise 54 An inventory number at Retro consists of three digits. The first two digits indicate product line and the next indicates style. Inventory is maintained on a computer system. If the operator mistypes one out of every 100 digits, what are the probabilities of the following?

(a) That when an item is sold, it will be incorrectly recorded.

(b) That when an item is sold, the wrong product line will be recorded regardless of style correctness.

(c) That the wrong product line and the correct style will be recorded.

(d) That the correct product line and wrong style will be recorded.

Exercise 55 Three out of 20 tyres are defective. If 5 of the 20 are randomly chosen for inspection, what is the probability that exactly two of the defective tyres will be chosen?

Exercise 56 There are six tennis teams. In how many ways can three teams finish first, second and third?

AN INTRODUCTION TO PROBABILITY

Exercise 57 A student can play squash for 0, 1 or 2 hours on any given night. Construct a tree diagram to determine the number of ways in which, in three nights, he can play squash for a total of

(a) 4 hours; and

(b) 5 hours.

Exercise 58 Determine the number of ways a student can mark her answers to a multiple choice test if there are:

(a) three questions with eight choices each; and

(b) ten questions with four choices each for the first three questions and five choices each for the next seven questions.

Exercise 59 Three out of 16 tax returns contain errors. If 7 returns are chosen randomly for audit, what is the probability:

(a) that exactly two of the returns with errors are audited; and

(b) that none of the returns with errors are audited?

Exercise 60 An accounting firm plans to hire two people from among five men and eight women. What is the probability that the first will be a man and the second a woman?

Exercise 61 A company estimates that the probability of a recession occurring next year is 0.4. The company also estimates the probability of an increase in interest rates in the next year at 0.5. Finally the company estimates the probability of both a recession and higher interest rates in the next year at 0.25

(a) If there is a recession, what is the probability of higher interest rates?

(b) What is the probability that there will be either a recession or higher interest rates or both?

Exercise 62 Two options available in a car are air-conditioning (C) and a radio (R). A dealer notes from sales records that the probability of a buyer selecting C is 0.6, and the probability of a buyer selecting R is 0.5. The probability that a buyer selects C if he selected R is 0.7. What is the probability:

(a) that a buyer selects both C and R; and

(b) that a buyer selects neither C nor R.

Exercise 63 A prospective buyer of a new car is faced with a choice of five body styles, three engine capacities and ten colours. How many different selections are open to him in choosing his new car?

Exercise 64   An urn contains 60 marbles: 40 are blue and 15 of these blue marbles are swirled. The rest of the marbles are red and ten of the red ones are swirled. The marbles that are not swirled are clear. What is the probability of drawing from the urn:
  (*a*) a red marble,
  (*b*) a clear marble,
  (*c*) a blue clear marble,
  (*d*) one blue swirled marble and one red swirled marble,
  (*e*) not a red clear marble.

Exercise 65   The personnel manager of a store wants to appoint three technikon students and two other people to fill five vacancies in his training programme.
  (*a*) In how many ways can these vacancies be filled if there are 12 technikon students among the 21 applicants.
  (*b*) What is the probability that three will be technikon students?

Exercise 66   Of 858 thefts committed in a certain city, 143 cases were never solved. What is the probability that a theft committed in this city will be solved?

Exercise 67   The Big Triple at a local racetrack consists of picking the correct order of finish of the first three horses in the ninth race. If there are 12 horses entered in the ninth race, how many possible Big Triple outcomes are there?

Exercise 68   Of 250 employees of a company, a total of 130 smoke cigarettes. Of the 150 male employees, 85 smoke cigarettes. What is the probability that an employee chosen at random:
  (*a*) does not smoke cigarettes;
  (*b*) is female and smokes cigarettes;
  (*d*) is male or smokes cigarettes.

Exercise 69   The probability that car sales will increase next month is estimated at 0.40. The probability that sales of spare parts will increase is estimated at 0.50. The probability that sales of both will increase is estimated at 0.10. What is the probability that:
  (*a*) car sales will increase if spare part sales have increased;
  (*b*) spare part sales will increase if car sales have increased;
  (*c*) car sales or spare part sales will increase.

Exercise 70   Of 100 individuals who applied for accounting positions with a large firm during the past year, 40 had some prior work experience and 30 had a professional certificate. However, 20 of the applicants had both work experience and a certificate. Determine the probability that a randomly selected candidate had:

(a) either work experience or a certificate;
(b) either work experience or a certificate but not both;
(c) a certificate, given that he had experience.

Exercise 71 A box contains 3 bags. Bag 1 contains 12 red, 4 green and 3 yellow marbles. Bag 2 contains 6 red, 2 green and 2 yellow marbles. Bag 3 contains 9 red, 7 green and 4 yellow marbles. If one marble is drawn from one of these bags, determine the probability that the marble will be red.

Exercise 72 An urn contains 8 red, 7 green and 5 white marbles. What is the probability that a marble selected at random will be:

(a) either red or green;
(b) either green or white.

Exercise 73 If one card is drawn from a pack of 52 cards, what is the probability that it will be a face card or a heart?

Exercise 74 An urn contains 14 red marbles and 6 green marbles. If sampling with replacement is done, what is the probability that a sample of two will contain:

(a) two red marbles;
(b) two green marbles.

Exercise 75 What is the probability of drawing an ace or a spade from a pack of playing cards?

Exercise 76 What is the probability of drawing a king or a queen from a pack of cards?

Exercise 77 If a die is rolled twice, what is the probability that the first roll yields a 5 or a 6 and the second roll anything but a 3?

Exercise 78 Hamo has a manager and two clerks. Whoever arrives first, opens the shop. Actually, one clerk arrives late 25 % of the time, the other clerk 10 % of the time and the manager 50 % of the time. If you arrive at the store at opening time, what is the probability that you will find it:

(a) open;
(b) open, but with only one person there.

Exercise 79 Suppose that the probability of getting an engaged tone when phoning a friend is 0.03. What is the probability of getting engaged tones in two calls to this friend on two different days?

Exercise 80 If the probability that a married man will vote in a certain election is 0.4 and the probability that a woman will vote provided that her husband votes, is 0.8, what is the probability that a husband and wife will both vote in this election?

Exercise 81 In a certain lottery, the probability of drawing a number divisible by two is $\frac{1}{2}$. The probability of drawing a number divisible by three is $\frac{1}{3}$ and the probability of drawing a number divisible by six is $\frac{1}{6}$. What is the probability of drawing a number which is divisible by either two or three?

CHAPTER 6
Probability Distributions

CONTENTS

6.1 INTRODUCTION 93
6.2 CLASSIFICATION OF PROBABILITY
    DISTRIBUTIONS 94
6.3 THE BINOMIAL PROBABILITY DISTRIBUTION 95
6.4 THE POISSON PROBABILITY DISTRIBUTION 96
6.5 THE NORMAL PROBABILITY DISTRIBUTION 97

6.1 INTRODUCTION

Probability distributions are related to frequency distributions. A frequency distribution is a listing of the observed frequencies of all the outcomes of an experiment. A probability distribution is a listing of all the possible outcomes that could result if the experiment were carried out, together with their probabilities.

Example   As an experiment, a coin is to be tossed three times.
(*a*) List of possible outcomes:
     HHH HHT HTH HTT THH THT TTH TTT
(*b*) A probability distribution showing the number of tails possible:

No. of Tails	Probability
0	$\frac{1}{8} = 0.125$
1	$\frac{3}{8} = 0.375$
2	$\frac{3}{8} = 0.375$
3	$\frac{1}{8} = 0.125$
	Total: $\frac{8}{8} = 1$

6.2 CLASSIFICATION OF PROBABILITY DISTRIBUTIONS

Probability distributions are classified as either discrete or continuous, depending on the random variable. A random variable is a numerical event whose value is determined by a chance process. For example: advertising expenditure is not a random variable because the level is chosen, whereas sales is not under the direct control of any observer and is therefore a random variable.

A discrete probability distribution is based on a discrete random variable which is one that can assume a countable number of numerical values. Example: For a sample of ten persons the number of persons favouring an increase in income tax can be only 0, 1, 2, 3, 4, 5, 6,7, 8, 9 or 10.

A continuous probability distribution is based on a continuous random variable which is one that can assume an infinitely large number of values within a given range.

Example   The outside diameter of a tree might be 6 m, 6.3 m, 6.327 m and so on, depending on the degree of accuracy of the tape measure being used.

An expected value of the random variable is a weighted average of the outcomes you expect in the future. To obtain the expected value, each value that the random variable can assume is multiplied by the probability of occurrence of that value, and then these products are summed.

$$E(x) = \sum x\, P(x)$$

Example   The following probability distribution together with the expected values has been computed for a company's yearly demand for cotton.

Demand (x) (kg)	Probability (P(x))	$\sum x\, P(x)$
3 000	0.2	600
4 000	0.4	1 600
4 500	0.2	900
5 000	0.2	1 000
		E(x): 4 100

There are many discrete probability distributions, but we will only focus on two: the binomial distribution and the Poisson distribution.

6.3 THE BINOMIAL PROBABILITY DISTRIBUTION

This is one of the probability distributions most widely used in determining the probability of occurrence where there are only two mutually exclusive outcomes to each trial — 'success' or 'failure'.

Characteristics
(1) Each trial has one of two possible mutually exclusive outcomes: 'success' or 'failure'.
(2) It is a discrete distribution resulting from the count of 'successes' in n trials, i.e. 0, 1, 2, ... n.
(3) Each trial is independent. This means that the outcome of one trial does not affect the outcome of any other trial.
(4) The probability of 'success' on each trial remains fixed from trial to trial.

$$P(x) = \frac{n!}{x!(n-x)!} \pi^x (1-\pi)^{n-x}$$

Where:
n = number of trials or sample size
π = probability of success on each trial
x = the binomial variable 0, 1, 2, ... n

Steps
❑ Determine the probability of a success for each trial (π).
❑ Determine the total number of trials (n).
❑ Decide on the number of successes for which you wish to determine the probability (x).
❑ Substitute the values into the formula.

The mean and standard deviation of the binomial distribution are:

Mean: $\mu = n\pi$
Standard deviation: $\sigma = \sqrt{n\pi(1-\pi)}$

Example The Surety Insurance Agency telephone is engaged 30 % of the time. Suppose five calls are made to Surety, what is the probability that one of the five callers will get an engaged signal? Determine the mean and standard deviation of the number of engaged signals.

$\pi = 0.3$
$n = 5$

$$x = 1$$
$$P(x=1) = \frac{n!}{x!(n-x)!} \cdot \pi^x \cdot (1-\pi)^{n-x}$$
$$= \frac{5!}{1!(5-1)!} \cdot 0.3^1 \cdot 0.7^4$$
$$= 0.3602$$

Determine the expected number of engaged signals, and the standard deviation.

$$E(x) = \mu = n\pi = 5 \times 0.3 = 1.5$$
$$\sigma = \sqrt{n\pi(1-\pi)} = \sqrt{5 \times 0.3 \times 0.7} = 1.0247$$

Using the same formula, the probability of any number of engaged signals can be determined, but it must be taken into account that the outcomes are mutually exclusive and that the special addition rule can therefore be applied if the probabilities of 'x or more' successes or 'x or fewer' successes are to be determined.
($x = 0$ or 1 or 2 or 3 or 4 or 5)

For example:
P(less than 2 engaged signals)
$= P(x = 0 \text{ or } 1) = P(x = 0) + P(x = 1)$

P(4 or more engaged signals)
$= P(x = 4 \text{ or } 5) = P(x = 4) + P(x = 5)$

P(at least 2 engaged signals)
$= P(x = 2 \text{ or } 3 \text{ or } 4 \text{ or } 5)$

If the number of individual probabilities to be determined is large, it is quicker to determine the complementary probabilities and subtract the sum from 1: $P(A) = 1 - P(\overline{A})$

For example
P(at least 2 engaged signals) $= 1 - [P(x = 0) + P(x = 1)]$

6.4 THE POISSON PROBABILITY DISTRIBUTION

This is a discrete probability distribution concerned with the number of successes that occur in a given unit of measure rather than the number of successes out of n trials.

Characteristics
(1) It is a discrete probability distribution
(2) The events are independent

PROBABILITY DISTRIBUTION

(3) There is only a single parameter μ, which is the mean number per unit of measure.
(4) The probability of an event occurring in all equal-sized units of measure is the same for all units.
(5) x is the count of the number of successes that occur in a given unit of measure and may take on any value from 0 to infinity.

$$P(x) = \frac{\mu^x \cdot e^{-\mu}}{x!}$$

Where:

x = value of the random variable, i.e. the number of occurrences in the given unit of measure.
μ = the average number of occurrences per unit of measure.
e = the base of the natural log. system (2.71828 ...)

If the unit of measure changes, the value of μ must change proportionately.

Example On average six people per hour use a self-service banking facility in a department store during prime shopping hours. What is the probability that:

- 3 people will use the facility during the next hour;
- 2 people will use the facility during the next ten minutes?

$$P(x = 3 \text{ if } \mu = 6 \text{ per hour}) = \frac{\mu^x \cdot e^{-u}}{x!}$$

$$= \frac{6^3 \cdot e^{-6}}{3!}$$

$$= 0.089$$

Changing the unit of time to 10 minutes changes μ from 6 people per 60 minutes to $\mu = 1$ per 10 minutes.

$$P(x = 2 \text{ if } \mu = 1) = \frac{1^2 \cdot e^{-1}}{2!}$$

$$= 0.1839$$

6.5 THE NORMAL PROBABILITY DISTRIBUTION

A continuous distribution is a distribution in which the x variable may assume any value within a given range or interval. The most widely used continuous probability distribution is the normal distribution.

Characteristics
(1) The graph is symmetrical and bell-shaped and the *x*-axis represents the possible values of the *x*-variable.
(2) The left and right hand tails of the distribution approach the *x*-axis, but never touch it.
(3) Probabilities for continuous variables correspond to areas under the normal curve.
(4) The normal distribution has two parameters: the mean and standard deviation. The centre of the distribution is determined by μ, and the spread of the distribution is determined by σ.
(5) The total area under the normal curve is equal to one or 100 %. This means that the areas to the right and the left of the mean will each comprise 50 % of the total area.
(6) It is possible to convert any normal distribution into a standard form (*z*), by expressing the differences between the value of interest (*x*) and the mean (μ) in units of standard deviation (*z*). This *z* will show the number of standard deviations that a particular value lies to the right or left of the mean. Any *x*-value smaller than the mean will have a negative *z*-score and any *x*-value greater than the mean will have a positive *z*-score.

$$z = \frac{x - \mu}{\sigma}$$

Normal distribution with a mean of 10, and a standard deviation of 2, converted into the standard normal distribution

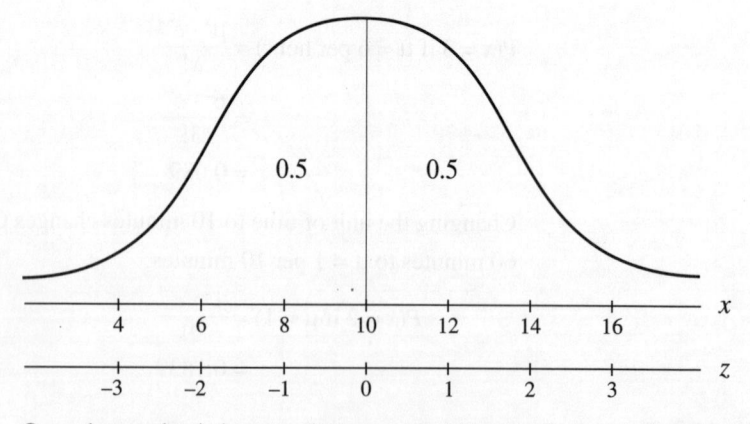

Once the *z*-value is known, the area under the curve between the mean and a *x*-value can be computed using the *z*-table (areas under the normal curve).

❑ If the area to be determined falls on both sides of the mean, determine the *z*-value for the area between the mean and the

x-value to the right of the mean. Determine a z-value for the area between the mean and the x-value to the left of the mean.

The z-value will be negative. Because of the symmetry of the normal curve, the area on the left between 0 and a negative z will be the same as the area on the right between 0 and z.

Obtain the areas for the two z-values from the z-table and add the values.

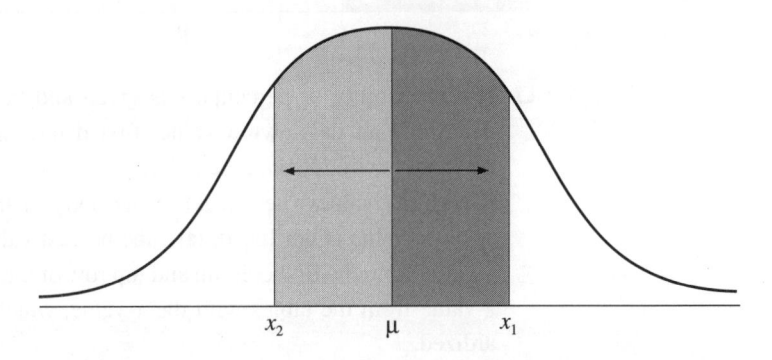

❏ If the area to be determined falls on one side of the mean, determine the z-value between the mean and the biggest x-value and also the z-value for the area between the mean and the smallest x-value.

Obtain the areas for the two z-values from the z-table and subtract the smaller value from the bigger one.

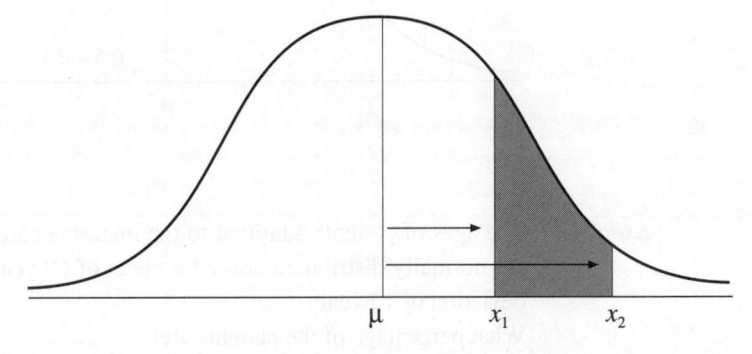

❏ If the area to be determined falls in one of the tails of the distribution, determine the z-value for the area between the mean and the x-value.

Obtain the area for the z-value from the z-table and subtract the value from 0.5.

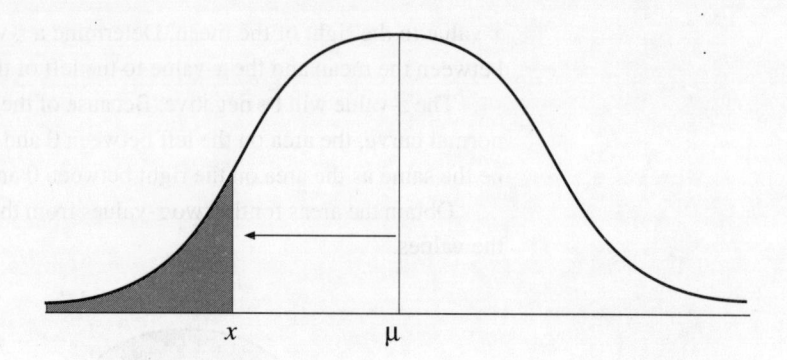

❏ If a probability or percentage is given and you are required to determine an unknown x-value, first determine the percentage between the middle and the unknown x-value, and then compare it with the values (i.e. areas) in the body of the z-table. If that specific value is not listed, take the nearest value and obtain the z-value from the first column and top row of the table. Equate this z-value from the tables with the x-value, which has been standardized.

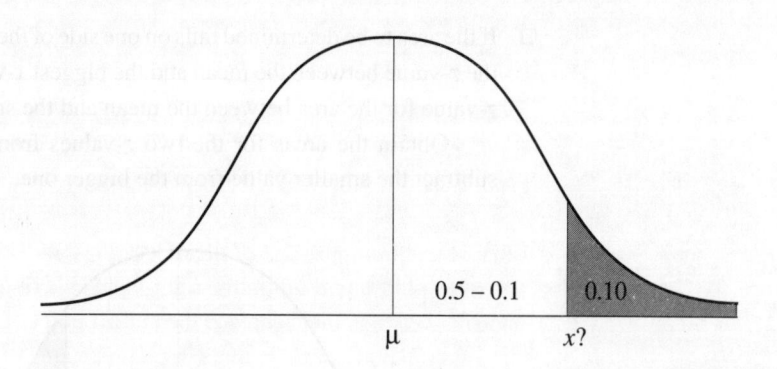

Example The ages of patients admitted to the intensive care unit of a hospital are normally distributed around a mean of 60 years with a standard deviation of 12 years.
What percentage of the patients are:

- *Between 78 and 45 years old.*

    The area of interest lies on both sides of the mean, therefore two areas above the intervals $(45 < x < 60)$ and $(60 < x < 78)$ need to be determined and then added.

$$z = \frac{x-\mu}{\sigma}$$

$$= \frac{78-60}{12} = 1.5 \quad \text{(area} = 0.4332)$$

$$z = \frac{45-60}{12} = -1.25 \quad \text{(area} = 0.3944)$$

$$P(45 < x < 78) = 0.4332 + 0.3944 = 0.8276 = 82.76\ \%$$

- *Older than 60 years*

  $P(x > 60) = 0.5 = 50\ \%$

- *Older than 78 years*

  $P(60 < x < 78) = 0.4332$
  $P(x > 78) \quad = 0.5 - 0.4332 = 0.0668 = 6.68\ \%$

- *Younger than 45 years*

  $P(45 < x < 60) = 0.3944$
  $P(x < 45) = 0.5 - 0.3944 = 0.1056 = 10.56\ \%$

- *Between 30 and 45 years*

  $$z = \frac{30-60}{12} = -2.5\ \text{(area} = 0.4938)$$

  $P(30 < x < 45) = 0.4938 - 0.3944 = 0.0994 = 9.94\ \%$

- What is the minimum age of the oldest 20 % of the patients?
  The oldest 20 % will fall in the right-hand side of the distribution. The area between the mean and the cut-off point of the oldest 20 % is 30 %. In the body of the z-table, look up an area of 0.3 or the nearest value to 0.3 and read off the z-value. Equate this to the standardized x-value.

  $$z = \frac{x-\mu}{\sigma}$$

  $$0.84 = \frac{x-60}{12}$$

  $x = 70.08$ years

Normal approximation to the binomial As the sample size n becomes larger, the shape of the binomial distribution becomes closer and closer to the shape of the normal distribution. Even though a formula exists for computing binomial probabilities exactly, the formula is simply not practical when dealing with large values of n.

A good rule to follow when deciding whether a sample size is big enough to apply the normal distribution to approximate the binomial, is to ensure that both $n\pi$ and $n(1 - \pi)$ exceed 5.

General procedure

Step 1   Determine n, the number of trials or sample size and π, the success probability.

Step 2   Ensure that both $n\pi$ and $n(1 - \pi)$ are at least equal to 5. If not, the normal approximation should not be used.

Step 3   Find π and σ using the formulae:
$$\mu = n\pi$$
$$\sigma = \sqrt{n\pi(1 - pi)}$$

Step 4   Make the correction for continuity. This correction needs to be done because the random variable in the binomial distribution is discrete whereas in the case of the normal distribution it is continuous. When using the normal curve areas to approximate the probability that a binomial random variable takes on a value between two integers, inclusive, subtract 0.5 from the smaller integer and add 0.5 to the larger integer before determining the area under the normal curve.

The continuity correction is only necessary when using the normal distribution to approximate probabilities for a discrete probability distribution, otherwise no continuity is required.

Step 5   Compute $z = \dfrac{x - \mu}{\sigma}$

Step 6   Look up the area concerned using the Normal Table.

Step 7   State the conclusion.

Example   The National Health Department reports that 40% of all injuries occur at home. Out of 500 randomly selected injuries, what is the probability that the number occurring at home is:
(1) exactly 190?
(2) between 180 and 210 inclusive?
(3) at most 225?
(4) more than 180? (excluding 180)

Step 1 $n = 500$ $\pi = 0.4$

Step 2 $n\pi = 500(0.4) = 200$
$n(1 - \pi) = 500(0.6) = 300$ (Both are > 5)

Step 3 $\mu = n\pi$
$= 200$
$\sigma = \sqrt{n\pi(1 - \pi)}$
$= \sqrt{500(0.4)(0.6)}$
$= 10.95$

(1) $P(x = 190) = P(189.5 \leq x \leq 190.5)$ with continuity correction.

$$z = \frac{x - \mu}{\sigma}$$

$$= \frac{189.5 - 200}{10.95}$$

$= -0.96$

Area: 0.3315

$$z = \frac{190.5 - 200}{10.95}$$

$= -0.87$

Area: 0.3078

$P(189.5 \leq x \leq 190.5) = 0.3315 - 0.3078$
$= 0.0237$

(2) $P(180 \leq x \leq 210) = P(179.5 \leq x \leq 210.5)$ with continuity correction.

$$z = \frac{179.5 - 200}{10.95}$$

$= -1.87$

Area: 0.4693

$$z = \frac{210.5 - 200}{10.95}$$

$= 0.96$

Area: 0.3315

$P(179.5 \leq x \leq 210.5) = 0.4693 + 0.3315$
$= 0.8008$

(3) $P(x \leq 225) = P(x \leq 225.5)$ with continuity correction.

$$z = \frac{225.5 - 200}{10.95}$$

$$= 2.33$$

Area: 0.4901

$$P(x \le 225.5) = 0.5 - 0.4901$$
$$= 0.0099$$

(4) $P(x > 180)$ doesn't change because 180 is excluded.

$$z = \frac{180 - 200}{10.95}$$
$$= -1.83$$

Area: 0.4664

$$P(x > 180) = 0.5 + 0.4664$$
$$= 0.9664$$

EXERCISES

Exercise 1 A hunter hits his target with 80 % of shots. If he shoots at the target 8 times, what is the probability that he will hit it no more than twice?

Exercise 2 Mr Zippo is late for 20 % of the classes he teaches. If the dean checks 10 of his classes, what is the probability that Zippo will be late for four or more of them?

Exercise 3 If 40 % of all patients have medical aid, what is the probability that in a sample of 10 patients:
(a) exactly 4 will have medical aid;
(b) at least 4 will have medical aid;
(c) 8 or more will have medical aid;
(d) no more than 1 will have medical aid.

Exercise 4 The probability of passing a given test is 0.9. If you take 4 tests, what is the probability that you will pass exactly 2?

Exercise 5 If actuarial tables indicate that 85 % of all 10-year olds will survive for 40 years, what is the probability that only one of three 10-year olds will survive for 40 years?

Exercise 6 The number of car accident insurance claims received by an insurance broker per day has a mean of one.
(a) On any given day, what is the probability that there will be no claims made?
(b) In any given 5-day week, what is the probability that there will be no claims made?

PROBABILITY DISTRIBUTION

(c) In a two-day period, what is the probability that exactly one claim will be made?

Exercise 7 By reviewing past records, it has been determined that a mean of two customers per hour arrive at J.C's Hair Village. What is the probability that in a two-hour period, exactly 8 customers will arrive?

Exercise 8 A telephone magazine sales person finds he averages 2 sales per hour of work. What is the probability that in a given hour of work he will make:
(a) exactly two sales?
(b) at least two sales?
(c) more than two sales?

Exercise 9 A particular all-round cricket player takes on average one wicket per innings. What is the probability that he will not take any wickets during the next two innings?

Exercise 10 The police records in Wynberg show that on a typical Friday between 10 p.m. and midnight, six persons are arrested for suspected drunken driving. It is 11p.m. on a Friday night. What is the probability that:
(a) for the next 20 minutes nobody will be arrested for suspected drunken driving?
(b) one person will be arrested for suspected drunken driving during the next 10 minutes?

Exercise 11 Research is being conducted on the use of cut flower preservative in order to study its effects on the vase life of the flower. It is known from past experience that the vase life of roses with preservative is normally distributed around a mean of 8 days with a standard deviation of 2 days. Determine the probability that the vase life of a rose with preservative will be:
(a) at most 4 days;
(b) between 6 and 9 days;
(c) between 9 and 11 days.
(d) Ten per cent of all cut roses with preservatives have a vase life less than how many days?

Exercise 12 A small town's records show that the daily maximum temperature during spring has averaged 15 °C with a standard deviation of 4 °C. Assume a normal distribution and determine the following:
(a) the percentage of days on which the maximum temperature will be between 15 °C and 19 °C;

(b) the percentage of days on which the maximum temperature will exceed 21 °C;

(c) the probability that the maximum temperature will be less than 12 °C;

(d) the probability that the maximum temperature will be between 13 °C and 14 °C;

(e) the maximum temperature that will be exceeded only 10 % of the time.

Exercise 13 A petrol distributor's records indicate that sales of super petrol average 3 500 litres per day with a standard deviation of 250 litres. Assuming that sales are normally distributed, determine the probability of daily sales between 3 300 and 3 800 litres.

Exercise 14 The daily demand for cheesecake at a restaurant is normally distributed with a standard deviation of 2.8. Determine the mean number of orders if less than 20 orders are received only 4 % of the time.

Exercise 15 A brewery filling machine is adjusted to fill bottles with 750 ml of beer, with a variance of 36. Periodically a bottle is checked and the amount of ale is noted. Assuming that the amount of fill is normally distributed, what is the probability that the next bottle checked contains more than 753 ml?

Exercise 16 The probability of ceramic bathtubs cracking in a kiln is 0.5. What is the probability that, in a randomly selected lot of eight tubs, four will crack?

Exercise 17 One quarter of all loans made by First Bank are made to clients who have had previous loans. What is the probability that, in a randomly selected set of five new loan accounts, exactly three will be obtained by clients who have had previous loans?

Exercise 18 A branch of National Bank is considering adding additional drive-in teller windows. It has found that cars arrive at such windows randomly at a rate of 2 cars every 6 minutes. What is the probability that:

(a) no cars arrive during a given 12-minute period?

(b) 3 cars will arrive during a given 3-minute period?

(c) more than one car will arrive during a given 15-minute period?

Exercise 19 A study has shown that 50 % of the houses in a certain area have swimming pools. Of five houses randomly selected in this area for a market research study, determine the probability:

(a) that two houses will have swimming pools; and

(b) that no more than one has a swimming pool.

Exercise 20 On average six people per hour use a self-service banking facility in a department store during prime shopping hours. What is the probability;

(a) that no one will use the facility during a given ten-minute period; and

(b) that at least one person will use the facility during a randomly selected hour?

Exercise 21 One third of all persons who have medical insurance with the Stover Agency make at least one claim during a year. What is the probability that three randomly selected accounts will have no claims this year?

Exercise 22 In a group of 200, the average score for a test was 50 with a standard deviation of 10. Assume a normal distribution.

(a) What proportion of the persons obtained a mark between 50 and 65?

(b) What percentage had marks above 65?

(c) How many persons obtained marks between 60 and 70?

(d) What is the chance that an unknown member of the group obtained a mark between 32 and 65?

Exercise 23 On the average, one out of every five boards purchased by a cabinet manufacturer is unusable for building cabinets. What is the probability that between 7 and 10 of a set of 12 boards are usable?

Exercise 24 Telephone calls come in at the rate of 3 per minute. What is the probability of exactly 5 calls arriving in a specified minute?

Exercise 25 It is known that 10 % of the items undergoing a certain process are damaged.

(a) In a sample of 15 items drawn at random, determine the expected number of damaged items and the probability of more than one damaged item.

(b) If the items are drawn one by one, determine the chance that the 12th item chosen will be the second damaged item.

Exercise 26 According to a technikon placement service, one out of every 5 job interviews on campus leads to a follow-up interview at the firm's premises. If 5 students each have an interview on campus, what is the probability that fewer than three will have follow-up interviews?

Exercise 27 Completion times for an assembly operation among a group of workers are normally distributed with a mean of 80 s and a standard deviation of 10 s. Determine the completion times that cut off the fastest 2.5 % and the slowest 2.5 %.

Exercise 28 A marketing research survey shows that approximately 80 % of the car owners surveyed indicated that their next car purchase would be either a compact or an economy car. If five prospective buyers are interviewed, determine the probability that:

(a) all five indicate that their next car will be either a compact or economy car;

(b) at most one indicates that his next purchase will be either a compact or an economy car.

Exercise 29 If an average of 40 service calls are required during a typical eight-hour shift in a manufacturing plant, what is the probability:

(a) that more than 10 service calls will be required during a particular hour; and

(b) that no service call will be required during a particular hour.

Exercise 30 The performance on a nationally standardized aptitude test in a certain professional field is reported in the form of transformed values having a mean of 500 with a standard deviation of 100. The scores are normally distributed.

(a) What percentage of the results are between 350 and 650?

(b) Out of 1 000 test results, how many would we expect to have a score of 700 or higher?

(c) Suppose that the top 10 % of the scores are regarded as being distinctly superior, what is the lowest score required to be included in this category?

Exercise 31 On the basis of past experience, truck inspector Charlie Ford has noticed that 7 % of all trucks coming in for their annual inspection fail the test. Using the normal approximation to the binomial distribution, find the probability that:

(a) between 10 and 20 of the next 200 trucks entering the inspection station will fail the inspection,

(b) exactly 50 of the next 200 trucks will pass the inspection.

CHAPTER 7
Statistical Estimation

CONTENTS

> 7.1 INTRODUCTION TO SAMPLING 109
> 7.2 SAMPLING DISTRIBUTIONS 110
> 7.3 ESTIMATION 113
> 7.4 INTERVAL ESTIMATES 113
> 7.5 DETERMINATION OF SAMPLE SIZE 119

7.1 INTRODUCTION TO SAMPLING

Statistical sampling is a systematic approach to selecting a few elements (sample) from an entire collection of data (population) in order to make some inferences about the total collection.

There are two methods of selecting samples from populations: non-probability sampling and probability or random sampling. The different methods of sampling were discussed in a previous chapter. The principles of simple random sampling are the basis of statistical inference; the process of making inferences about populations from information contained in samples. The characteristics of the population or samples are the measures by which they are described. These would include the arithmetic mean, the standard deviation, the size of the group, etc. If a particular measure describes a population, it is called a parameter. If a particular measure describes a sample, it is called a statistic or estimator.

Parameters	*Statistics*
μ = mean	$\bar{x}$ = mean
σ = standard deviation	s = standard deviation
N = population size	n = sample size
π = population proportion	p = sample proportion

7.2 SAMPLING DISTRIBUTIONS

The values of a statistic, such as the mean, calculated from all the possible samples of given size n from a population will differ from one sample to the next. These differences are the result of chance errors. When all the sample means that may occur are organized into a probability distribution, the resulting distribution is called the sampling distribution of the mean.

If the mean of the population distribution is μ, and its standard deviation is σ, the distribution of the sample means will have a mean of: $\mu_{\bar{x}} = \mu$
and a standard deviation of: $\sigma_{\bar{x}} = \dfrac{\sigma}{\sqrt{n}}$

The standard deviation of the sampling distribution of the mean is known as the standard error of the mean, because it measures the extent to which the means from the different samples are expected to vary due to chance error in the sampling process. The standard error indicates not only the size of the chance error, but also the likely accuracy if a sample statistic is used to estimate a population parameter. In most cases the population standard deviation is unknown. However, if the sample size is 30 or more, the sample standard deviation (s) is a good estimate of the population standard deviation (σ). The standard error of the mean then becomes:

$$s_{\bar{x}} = \dfrac{s}{\sqrt{n}}$$

Characteristics
(1) When the population is normally distributed, the sampling distribution of the mean is also normal, regardless of the sample size.
(2) The central limit theorem
The use of the z-table for calculating probabilities does require a normally distributed population. Since most distributions are not exactly normally distributed, probabilities of this type cannot be determined without a method of dealing with non-normally distributed populations. The central limit theorem deals with such situations:
   If the population from which the sample is drawn is not normal, the distribution of the sample means will become more and more normal as the sample size increases. A sample of 30 or more is considered by most as being large enough. The significance of this theorem is that it permits the use of the sample mean to make inferences about the population mean without knowledge

of the shape of the probability distribution of the original population.

(3) When the population standard deviation (σ) is unknown, the sample standard deviation (s) may provide a poor estimate of σ when the sample size is small ($n < 30$). Instead of using the normal distribution which requires σ or a good estimate of σ, we use the t-distribution, provided that the population we are sampling from has the shape of the normal distribution. The standard error of estimate:

$$s_{\bar{x}} = \frac{s}{\sqrt{n}}$$

The t-distribution is a symmetrical distribution. For relatively small samples, it is much flatter than the normal distribution, but as the sample size increases, the t-distribution becomes more and more normal. Tables of t-values are usually only completely enumerated for $(n - 1) \leq 30$, because for larger samples the normal distribution gives a very good approximation of the t-values and is easier to use.

There are infinitely many t-curves, and the one that is used depends on the quantity $(n - 1)$ — sample size less 1 — called the number of degrees of freedom (df)

The mathematical concepts involved in defining degrees of freedom are somewhat complex. Therefore we define degrees of freedom simply as a number that identifies the appropriate t-distribution.

(4) Many populations are finite, that is, of a stated or limited size, for example the number of students in a class, a day's production, etc. In a finite population the probability that an individual item will be chosen increases significantly as the sample becomes larger and larger. For example, if a sample of $n = 10$ is to be selected from a population of $N = 100$, the first item in the sample has a $1/100$ probability of being selected. The tenth item has a $1/91$ probability of being selected.

To deal with this problem we need to modify our standard error of infinite populations by multiplying the original standard error by a finite population correction factor.

The correction factor is:

$$\sqrt{\frac{N-n}{N-1}}$$

and the standard error of the means becomes:

$$\sigma_{\bar{x}} = \frac{\sigma}{\sqrt{n}} \cdot \sqrt{\frac{N-n}{N-1}}$$

This correction is negligible if the sample is but a small proportion of the total population of items. The rule of thumb we use is that the finite correction factor may be omitted when the sample size is less than 5 % of the population size.

(5) Sometimes the parameter of interest is not the arithmetic mean, but a proportion (π). The way in which sample proportions (p) are distributed (where p is defined as the ratio of the number of successes to the total number of events in a binomial situation) must therefore be investigated. The sampling distribution of proportions is approximately normal if $np \geq 5$ and $nq \geq 5$, with a standard error of:

$$\sigma_p = \sqrt{\frac{\pi(1-\pi)}{n}}$$

If π is unknown, the estimated standard error of the proportion is:

$$s_p = \sqrt{\frac{pq}{n}} \quad \text{(infinite proportions)}$$

$$\text{Note: } q = 1 - p$$

$$s_p = \sqrt{\frac{pq}{n}} \cdot \sqrt{\frac{N-n}{N-1}} \quad \text{(finite proportions)}$$

(6) When the sample is small ($n < 30$) and the population cannot be assumed to be normally distributed, neither the normal distribution nor the t-distribution can be used for estimation.

Some distribution-free techniques known as 'non-parametric' statistical tests can be used, but these are not dealt with in this book.

7.3 ESTIMATION

Statistical inference is based on estimation and hypothesis testing (the subject of the next chapter). In both estimation and hypothesis testing, inferences about population parameters are made from information contained in samples.

Two types of estimates can be made about a population: a point estimate and an interval estimate. A point estimate is a single number or statistic taken as an estimate of an unknown population parameter by itself. It is often insufficient because its reliability is uncertain. In order to make a point estimate of the population mean (μ), a sample of size n is taken, and the sample mean ($\bar{x}$) is computed. The sample mean ($\bar{x}$) is a point estimate of the population mean (μ).

A point estimate is much more useful if it is accompanied by an estimate of the error that might be involved — i.e. an interval estimate. An interval estimate is an estimate of the range of values within which the population parameter may, with some confidence, be expected to lie. The width of the interval will depend on the sample data used in making a point estimate, on the sample size, the variation in the data and the degree of accuracy required.

7.4 INTERVAL ESTIMATES

The probability associated with an interval estimate is called the confidence level, and the interval estimate is called a confidence interval. This probability is a measure of the confidence that the interval estimate will include the population parameter. A high probability, say 99 %, means more confidence and hence a wider interval. However, if the level of confidence is too high the confidence interval will be too wide and the estimate may have little real value. If the confidence interval is too narrow on the other hand, the estimate is associated with such low confidence that its value is questioned.

In practice, it is desirable to construct an interval that has a reasonably high confidence level and is as small as possible. An increase in sample size will decrease the width of the interval for a given confidence level.

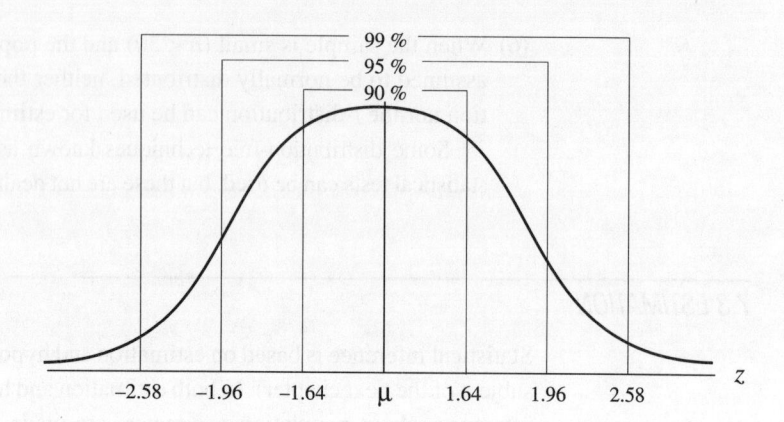

Steps in calculating a confidence interval

(1) Collect a sample of size n. Some sampling error will arise because the whole population has not been studied. This error is controlled by selecting a sample of adequate size (refer to page 120).
(2) Determine the type of sampling distribution.

Parameter to be estimated	Population distribution	Population standard deviation (σ)	Sample size (n) (Note 1)	Sampling distribution	Formulae (Note 2)
μ	Normal	Known	Large	Normal (z)	$\mu = \bar{x} \pm z \cdot \dfrac{\sigma}{\sqrt{n}}$
			Small		
	Unknown	Known	Large	Normal (z) via central limit theorem	
			Small	—	—
	Normal	Unknown	Small	t-distribution	$\mu = \bar{x} \pm t \cdot \dfrac{s}{\sqrt{n}}$
			Large	Normal (z)	$\mu = \bar{x} \pm z \cdot \dfrac{s}{\sqrt{n}}$
	Unknown	Unknown	Large	Normal (z) via central limit theorem	
			Small	—	—
π	Normal	—	Large	Normal (z)	$\pi = p \pm z \cdot \sqrt{\dfrac{p(1-p)}{n}}$
			Small		
	Unknown	—	Large	Normal (z) if both $n\pi$ and $n(1-\pi) > 5$	
			Small	f-distribution	—

STATISTICAL ESTIMATION

Note 1:

Large sample: $n \geq 30$ Small sample: $n < 30$

Note 2:

For finite populations, multiply standard error of infinite populations formulae by finite correction factor: $\sqrt{\dfrac{N-n}{N-1}}$

---

(3) Compute the sample mean ($\bar{x}$) and standard deviation (s) or proportion (p).

(4) Determine the value of z or t by making use of the appropriate table and the degree of confidence required.

(5) Determine the standard error of estimate.
Is the population finite or infinite? Remember to make use of the finite correction factor when computing the standard error of estimate.

(6) Calculate the upper and lower confidence limits by making use of the appropriate formula.

(7) Briefly state the meaning of your confidence interval: the probability that the interval will contain μ is 95 % or 90 % etc., depending on the confidence level.

Examples: Population σ known

Example 1   The Pappi Paper Company wants to estimate the average time required for a new machine to produce a ream of paper. A sample of 36 reams required an average production time of 1.5 min for each ream. Assuming $\sigma = 0.30$ min, construct an interval estimate with a confidence level of 95 %.

$$\mu = \bar{x} \pm z \cdot \frac{\sigma}{\sqrt{n}}$$

$$= 1.5 \pm 1.96 \cdot \frac{0.3}{\sqrt{36}}$$

$$= 1.5 \pm 0.098$$

$$\approx 1.402 \leq \mu \leq 1.598$$

Interpretation

The Pappi Paper Company can be 95 % sure that the average time required for a new machine to produce a ream of paper lies between 1.402 and 1.598 minutes.

You can make this statement, because in repeated sampling, 95 % of the intervals you construct in this manner will contain the population mean.

Example 2 The Longmore Company has received a shipment of 100 lengths of pipe, and wants to estimate the average diameter of the pipes to see if they meet minimum requirements. A sample of 50 pipes had an average diameter of 255 mm. In the past the population standard deviation of the diameter has been 7 mm. Construct an interval estimate with a 99 % degree of confidence.

$$\mu = \bar{x} \pm z \cdot \frac{\sigma}{\sqrt{n}} \cdot \sqrt{\frac{N-n}{N-1}}$$

$$= 255 \pm 2.58 \cdot \frac{7}{\sqrt{50}} \cdot \sqrt{\frac{100-50}{100-1}}$$

$$= 255 \pm 1.82$$

$$\approx 253.18 \leq \mu \leq 256.82$$

The Longmore Company can be 99 % confident that the average diameter for all 100 lengths of pipe is between 253,18 and 256,82 mm.

Population σ unknown

Example 1 The Bow-Tie Tavern wants to estimate the average spending per customer. A sample of 100 customers spent an average of R3.50 each with a sample standard deviation of R0.75. Estimate the true average expenditure with a 90 % confidence level.

For this sample $n > 30$, therefore the *s*-value will give a good approximation of the σ-value.

$$\mu = \bar{x} \pm z \cdot \frac{s}{\sqrt{n}}$$

$$= 3.50 \pm 1.64 \cdot \frac{0.75}{\sqrt{100}}$$

$$= 3.50 \pm 0.123$$

$$\approx 3.377 \leq \mu \leq 3.623$$

$$R3.38 \leq \mu \leq R3.62$$

STATISTICAL ESTIMATION

The Bow-Tie Tavern can be 90 % confident that the average spending per customer is between R3,38 and R3,62.

Example 2 The Hen-Inn Company has received a shipment of 100 hens, and the manager wants to estimate the true average mass of a hen in order to determine if the hens meet Hen-Inn's standards. A sample of 36 hens has shown an average mass of 3.6 kg with a sample standard deviation of 0.6 kg. Cconstruct an interval estimate of the true average mass per hen with a 99 % confidence level.

$$\mu = \bar{x} \pm z \cdot \frac{s}{\sqrt{n}} \cdot \sqrt{\frac{N-n}{N-1}}$$

$$= 3.6 \pm 2.58 \cdot \frac{0.6}{\sqrt{36}} \cdot \sqrt{\frac{100-36}{100-1}}$$

$$= 3.6 \pm 0.21$$

$$\approx 3.39 \le \mu \le 3.81$$

The Hen-Inn Company can be 99 % confident that the average mass for all 100 hens is between 3.39 kg and 3.81 kg.

Example 3 The Milky Inn wants to estimate the average number of litres of product sold per day. Twenty business days were monitored, and an average of 32 litres was sold daily with a standard deviation of 12 litres. Assume a normal distribution and calculate the confidence limits at the 95 % confidence level.

$$\mu = \bar{x} \pm t \cdot \frac{s}{\sqrt{n}}$$

$$= 32 \pm 2.093 \cdot \frac{12}{\sqrt{20}}$$

$$= 32 \pm 5.616$$

$$\approx 26.38 \le \mu \le 37.616$$

The Milky Inn can be 95 % confident that the average quantity sold per day is between 26,38 and 37,616 litres.

Example 4 The Kite Doe Company wants to estimate its average daily usage of flour per month. In a sample of 14 days, the mean flour usage was calculated as 173 kg with a standard deviation of 45 kg. Assume the distribution is normal and construct a confidence interval with a 99 % confidence coefficient.

$$\mu = \bar{x} \pm t \cdot \frac{s}{\sqrt{n}}$$

$$= 173 \pm 3.012 \cdot \frac{45}{\sqrt{14}}$$

$$= 173 \pm 36.22$$

$$= 136.78 \leq \mu \leq 209.22$$

The Kite Doe Company can be 99 % confident that the mean daily flour usage is between 136,78 and 209,22 kg.

Population percentage

Example 1   The Tug-Bolt Company wants to estimate the percentage of credit customers who have submitted cheques in payment for bolts and whose cheques have bounced. A sample of 150 accounts showed that 15 customers have passed bad cheques. Estimate at the 95 % confidence level the true percentage of credit customers who have passed bad cheques.

$$p = \frac{15}{150} = 0.1 \text{ and } q = 1 - 0.1 = 0.9$$

Use the z-table because $np \geq 5$ and $nq \geq 5$.

$$\pi = p \pm z \cdot \sqrt{\frac{pq}{n}}$$

$$= 0.1 \pm 1.96 \cdot \sqrt{\frac{0.1 \times 0.9}{150}}$$

$$= 0.1 \pm 0.048$$

$$\approx 0.052 \leq \pi \leq 0.148 \approx 5.2\% < \pi < 14.8\%$$

The Tug-Bolt Company can be 95 % confident that the number of customers who submit bad cheques is between 5.2 % and 14.8 %.

Example 2   A student counsellor was interested in the proportion of male students who would volunteer for military service. Out of 600 male students, she sampled 50 and found that 15 of them volunteered. Use a 99 % confidence coefficient to estimate the true percentage.

$$p = \frac{15}{50} = 0.3 \text{ and } q = 1 - 0.3 = 0.7$$

$$\pi = p \pm z \cdot \sqrt{\frac{pq}{n}} \cdot \sqrt{\frac{N-n}{N-1}}$$

$$= 0.3 \pm 2.58 \cdot \sqrt{\frac{0.3 \times 0.7}{50}} \cdot \sqrt{\frac{600-50}{600-1}}$$

$$= 0.3 \pm 0.16$$

STATISTICAL ESTIMATION

$$\approx 0.14 \leq \pi \leq 0.46$$
$$\approx 14\% \leq \pi \leq 46\%$$

The student counsellor can be 99 % confident that the proportion of male students who would volunteer for military service is between 14 % and 46 %.

Population percentage: small samples (optional) This approach is usually limited to small samples ($np < 5$) and uses the F-distribution (which is discussed in more detail in chapter 10).

Example A random sample of 27 electric toasters from a production line has been tested and three have been found to be defective. A 90 % confidence interval estimate of the process proportion defective is required.

Lower limit: $$L = \frac{x}{x + (n - x + 1)F_1}$$

Upper limit: $$U = \frac{(x + 1)F_2}{(x + 1)F_2 + (n - x)}$$

where: $$F_1 = F\left[1 - \frac{\alpha}{2}; 2(n - x + 1) \cdot 2x\right]$$

$$F_2 = F\left[1 - \frac{\alpha}{2}; 2(x + 1), 2(n - x)\right]$$

$$L = \frac{3}{3 + (27 - 3 + 1)3.75} = 0.03 \quad \text{where:}$$

$$F_1 = F[0.95; 50, 6] = 3.75 \quad \text{(F–distribution table)}$$

$$U = \frac{4(2.14)}{4(2.14) + 24} = 0.26 \quad \text{where:}$$

$$F_2 = [0.95; 8, 48] = 2.14 \quad \text{(F–distribution table)}$$

The 90 % confidence interval estimate is:

$$0.03 < \pi < 0.26$$
$$3\% < \pi < 26\%$$

7.5 DETERMINATION OF SAMPLE SIZE

In practice the desired interval length and confidence level are often specified, and it is necessary to determine the sample size to meet these specifications. Sample size varies inversely to the interval length: the larger the sample, the shorter the interval for a given confidence level. If a sample is too large, collecting the data would be a waste of money and effort. Similarly, if the sample is too small, the resulting conclusion will be uncertain. The correct sample size depends on three factors:

(1) the level of confidence desired;
(2) the variability in the population being studied; and
(3) the maximum allowable error.

The researcher selects the level of confidence. If the population is widely dispersed, a large sample is required, while a small standard deviation would not require as large a sample. The maximum allowable error (E) is the amount that was added and subtracted from the sample mean to obtain the limits of the confidence interval. A small allowable error will require a larger sample size, whereas a large allowable error will require smaller sample sizes.

Sample size for estimating a mean

$$n = \left(\frac{z \cdot \sigma}{E}\right)^2$$

Steps in determining the sample size

- Determine the level of confidence desired.
- Determine the maximum allowable error, i.e. the difference between the sample mean (x) and the population mean (μ).
- Find the standard deviation.
- Substitute into the equation.
- Since the result of this computation is not always a whole number, the usual practice is to round up any fractional result.

Example Consider a machine that is filling cans with tomato paste. Experience has shown that in this process the population of fill masses are normally distributed with a standard deviation of 0.31 g. The production supervisor wants to collect a sample just large enough to provide a sample mean within 0.25 g of the true process mean at the 99 % confidence level.

$$n = \left(\frac{z \cdot \sigma}{E}\right)^2$$

$$= \left(\frac{2.58 \cdot 0.31}{0.25}\right)^2$$

$$= 10.23$$

$$\approx 11$$

Sample size for estimating a proportion

It is possible to determine the sample size necessary to provide a sample proportion (*p*) that is within a certain distance of the true population proportion (π).

$$n = \frac{\pi(1-\pi)z^2}{E^2}$$

Example A political candidate has employed a consultant famous for polls. He asked him to determine within 3 % the proportion of people who are currently likely to vote for him. The level of confidence desired is 95 %. Assume the candidate's previous popularity to be 50 %, and determine the correct sample size.

$$n = \frac{\pi(1-\pi)z^2}{E^2}$$
$$= \frac{0.5(0.5)1.96^2}{0.03^2} = 1\,067$$

EXERCISES

Exercise 1 A sample of 900 registered voters were surveyed about their age. The mean of the sample was computed to be 42 years and the standard deviation 12 years. What is the 95 % confidence interval for the population mean?

Exercise 2 A simple random sample of size 300 is taken from a population of clerical workers, and the data on their hourly wages are recorded. The sample mean and the standard deviation are found to be R4.44 and R1.27 respectively. Construct a 90 % confidence interval for the mean hourly wage of this population.

Exercise 3 A random sample of 80 school children enrolled at Home Street School were asked how far they travel to school. The sample mean and standard deviation were found to be 7.52 km and 1.32 km respectively. What is the 95 % confidence interval for the population mean distance traveled to school?

Exercise 4 Information with respect to income, number of children, their ages, and so on was obtained from a sample of 160 families. 'What is the age of your youngest child?' was one of the questions asked. The sample mean was computed to be 6.7 years with a standard deviation of 2.5 years. Construct a 99 % confidence interval for the mean age of the youngest child in all families.

Exercise 5 A study is to be conducted on the mean salary of mayors of towns with populations of less than 100 000. The error in estimating the mean is to be less than R100, and a confidence level of 95 % is desired. Suppose the standard deviation of the population is estimated to be R1 000, what is the required sample size?

Exercise 6 A company wishes to estimate the average starting salaries for security personnel in their factories. From a previous study they estimate the standard deviation at R2.50. How large a sample should be selected in order to be 95 % confident that the sample mean differs from the population mean by at most R0.50?

Exercise 7 A study is conducted with a view to estimating the mean number of hours worked per week by suburban housewives. The pilot study revealed that the population standard deviation is 2.7 hours. How large a sample should be selected to obtain 95 % confidence that the sample mean differs from the population mean by at most 0.2 hours?

Exercise 8 A meat packer is investigating the marked mass shown on vienna sausages. A pilot study showed a mean mass of 11.8 kg per pack and a standard deviation of 0.7 kg. How many packs should be sampled in order to be 95 % confident that the sample mean differs from the population mean by at most 0.2 kg?

Exercise 9 The owner of Power Flower desires to estimate the mean size of her outstanding accounts receivable. A random sample of 25 accounts from a normal population yielded an average unpaid balance of R60 with a standard deviation of R10. Construct a 90 % confidence interval estimate of the population mean.

Exercise 10 The tear strength of a particular paper product is known to be normally distributed. If a random sample of 9 rolls yielded a mean tear strength of 225 kg/m^2 with a standard deviation of 15 kg/m^2, construct a 90 % confidence interval estimate for the average tear strength.

Exercise 11 The number of shirts finished per hour by a particular production line is normally distributed. A random sample of 25 hours output had a mean of 40 shirts per hour and a standard deviation of 9 shirts. Construct a 95 % confidence interval estimate of the mean number of shirts produced per hour.

STATISTICAL ESTIMATION

Exercise 12 A random sample of 25 motels revealed that the average room charge was R45 per day with a standard deviation of R6. Assuming that charges are normally distributed, construct a 90 % confidence interval estimate for μ.

Exercise 13 A random sample of 100 families contained 65 families that owned one or more colour television sets. Based on the sample results, construct a 99 % confidence interval estimate of the proportion of the population that owns colour television sets.

Exercise 14 In a physical fitness test 240 out of a random sample of 400 high school students could run a mile in less than 8 minutes. Construct a 90 % confidence interval estimate of the proportion of all high school students able to run a mile in less than 8 minutes.

Exercise 15 A large bank recently studied a random sample of 80 accounts and observed that 20 of the account-holders would not have to pay a service charge if a new minimum balance policy was adopted. Use the sample result to construct a 90 % confidence interval estimate of the proportion of bank customers that would benefit from the new policy.

Exercise 16 Repeat exercise 15, but now based on 600 customers that might be affected by the new policy.

Exercise 17 A morning newspaper is investigating the reading habits of the home delivery customers. If a previous survey indicated that 50 % of readers always 'look over' the editorial page, what sample size should be used to estimate within 4 % the proportion of people who always 'look over' the editorial page. Set $\alpha = 0.10$.

Exercise 18 Mr Bell would like to estimate the proportion of families who prefer push-button phones. If 40 % of the families were known previously to prefer push-button phones, what size simple random sample is needed to estimate within 4 % of the actual proportion the current proportion who prefer push-button phones? Set $\alpha = 0.10$

Exercise 19 If a simple random sample of residents is to be taken to determine how they expect to vote on a proposal to increase the property tax, what sample size would be necessary to keep the sample proportion within 3 % of the population proportion? Assume a 50–50 split and $\alpha = 0.05$.

Exercise 20 A magazine publishing company wants to estimate the proportion of its customers that would purchase television programme guides. The company requires 95 % confidence that its estimate is correct to within

0.05 of the true proportion. Past experience in other areas indicates that 30 % of the customers will purchase the programme guide. What sample size is needed?

Exercise 21 A recent random sample of 36 car owners revealed an average weekly expenditure on petrol of R30 with a standard deviation of R12. Construct a 95 % confidence interval estimate for the population mean expenditure.

Exercise 22 A random sample of 49 pocket calculators yielded a mean life of 21 months with a standard deviation of 2.8 months. Construct a 95 % confidence interval estimate for the average life expectancy.

Exercise 23 A manufacturer of a certain type of X-ray developing fluid has recently collected information on the life expectancy of the fluid from a random sample of 64 customers. The random sample yielded a sample mean of 16 days and a standard deviation of 4 days. Construct a 90 % confidence interval for the average life expectancy of the fluid.

Exercise 24 A random sample of 16 boxes of a popular cereal contained a mean of 100 raisins per box with a standard deviation of 36 raisins per box. Assume that the number of raisins per box is normally distributed, and construct a 95 % confidence interval estimate.

Exercise 25 Refer to exercise 24. The machine putting the raisins in the cereal boxes went out of adjustment and 400 boxes were processed before the error was corrected. Estimate, with a 95 % confidence interval, the mean number of raisins per box in this batch if the sample of 16 was taken from it.

Exercise 26 The manufacturer of a steel product wishes to estimate the proportion of defective items manufactured. A random sample of 200 items produced 10 defectives. Construct a 99 % confidence interval estimate for π.

Exercise 27 A manufacturer of bicycles recently observed the performance of a random sample of 50 bicycles. Fifteen of the bicycles were observed to perform improperly on the hill climb. Construct a 99 % confidence interval estimate of the proportion of bicycles having this malfunction.

Exercise 28 A survey is planned to determine the average annual family medical expenses of employees of a large company. The management of the company wishes to be 95 % confident that the sample average is correct to within ±R50 of the true average family expenses. A pilot

STATISTICAL ESTIMATION

study indicated that the standard deviation can be estimated at R400. How large a sample size is necessary?

Exercise 29 The mean current ratio of 100 firms selected at random from a total of 5 000 shoe manufacturing businesses is 1.2 with a standard deviation of 0.9. Determine the 95 % confidence limits for the true mean current ratios of the 5 000 firms.

Exercise 30 Trans Trucking Company charges for delivery by the mass of the cargo and the distance conveyed. Determine the sample size that would be necessary to estimate the mass of full truck delivery between two cities if we desire our estimate to be accurate to within 10 kg with 95 % confidence. Assume that the standard deviation of such masses is 30 kg.

Exercise 31 Sixteen employees of a telephone company were selected in a random sample to determine their average monthly telephone bill. The mean monthly bill was found to be R28 with a standard deviation of R5. Construct a 98 % confidence interval for the average monthly bill for all of the company's employees. Assume the distribution of telephone bills is normal.

CHAPTER 8
Hypothesis Testing for Population Means and Proportions

CONTENTS

8.1 INTRODUCTION 127
8.2 GENERAL STEPS IN HYPOTHESIS TESTING 128
8.3 INFERENCES INVOLVING TWO POPULATIONS . . 138
8.4 HYPOTHESIS TESTING BASED ON PAIRED
    OBSERVATIONS 145

8.1 INTRODUCTION

A hypothesis is an assumption to be tested with the objective of making statistical decisions based on a scientific procedure. It is used when attempting to determine when it would be reasonable to conclude, from analysis of a sample, that the entire population possesses a certain property. A parameter value is hypothesized, and sample data are then collected to determine whether the hypothesized value should be rejected or accepted as being correct.

We may find that the two values differ, and the question we need to answer is whether the difference between the hypothesized population mean (μ_0) and the sample mean ($\bar{x}$) is significant and therefore not supportive of the hypothesis, or whether the difference is due to chance and therefore supportive of the hypothesis.

When we draw a sample, any one of the samples in the sampling distribution might be selected. Most of the sample means would not equal the population mean. Such a difference between a sample mean and a population mean is due to the sampling process and is known as a chance difference.

A statistically significant difference is a difference just too great to be attributed to chance. Such a real difference between the sample mean and the population mean indicates that the sample with mean $\bar{x}$ comes from a population with a mean other than the hypothesized mean (μ_0).

8.2 GENERAL STEPS IN HYPOTHESIS TESTING

(1) State the null hypothesis H_0:

The null hypothesis is a claim about the population parameter to be tested. It can be rejected or not rejected. The test of H_0 can be either one- or two-sided, depending on whether deviations in only one or in both directions are being investigated.

$H_0: \mu = \mu_0$; $H_0: \mu \leq \mu_0$: $H_0: \mu \geq \mu_0$ where μ_0 is the hypothesized value.

(2) State the alternative hypothesis H_a:

If sample results fail to support H_0, it must be concluded that some other hypothesis is true. H_a is a claim about the population parameter that is accepted if H_0 is rejected. The following possibilities exist:

If $H_0: \mu = \mu_0$; then $H_a: \mu \neq \mu_0$; (two-tail) or
$\qquad\qquad\qquad\qquad\quad \mu \leq \mu_0$; (one–tail test to the left)
$\qquad\qquad\qquad\qquad\quad \mu \geq \mu_0$; (one–tail test to the right)
If $H_0: \mu \geq \mu_0$; then $H_a: \mu < \mu_0$; (one-tail)
If $H_0: \mu \leq \mu_0$; then $H_a: \mu > \mu_0$; (one-tail)

A test in which we want to find out whether a population parameter (μ or π) has changed, regardless of the direction of change, is referred to as a two-tail test.

If we wish to determine whether the sample came from a population that has a parameter (μ or π) less than a hypothesized value or from a population that has a parameter more than a hypothesized value, the attention is focused on the direction of change, and the test is referred to as one-tail.

(3) Specify the level of significance to be used (α).

The purpose of the level of significance (α) is to provide a basis for deciding whether an observed difference between a sample statistic and a hypothesized parameter is a chance difference or a statistically significant difference. The significance level is the probability level at

HYPOTHESIS TESTING

which the decision-maker concludes that an observed difference cannot be atributed to chance.

There is a risk, however, that our conclusion will be wrong, because sampling is used and sampling is subject to error. Two incorrect decisions can be made in hypothesis testing:

(1) rejection of a true null hypothesis, or

(2) failure to reject a false null hypothesis.

Such errors are referred to as, respectively, type 1 and type 2 errors. The level of significance (α) refers to the maximum probability of making a type 1 error. How does one decide what level of significance to select?

If rejecting a true null hypothesis is relatively serious, set the significance level quite low. If accepting a false null hypothesis is relatively more serious, then set the significance level high. Testing procedures are designed to balance the risk of making either type of error. By choosing a smaller α, the chance of rejecting a true null hypothesis will be smaller, with a resulting greater chance of accepting a false null hypothesis, i.e., a greater β risk.

Usually tests are performed at a level of significance of 0.01, 0.02, 0.05 or 0.10.

(4) Select the test statistic (z or t) to be used by determining the type of sampling distribution.

A test statistic is a statistic based on a sample that is used in hypothesis testing to measure how close the sample statistic has come to the hypothesized value in H_0. The sample statistic (i.e. $\bar{x}$ and p) is converted in a standard deviation unit z or t, which is then used as the test statistic.

$$\text{test statistic} = \frac{\text{sample statistic} - \text{hypothesized parameter}}{\text{standard error of the statistic}}$$

A test statistic may assume many possible values when H_0 is true — one for each possible sample in the sampling distribution. This distribution of test statistics is known as the sampling distribution of the test statistic.

Identify the appropriate sampling distribution of the test statistic (i.e. normal z- or t-distribution) by making use of the following table:

Parameter to be hypothesized	Population distribution	Population standard deviation (σ)	Sample size (n) (Note 1)	Sampling distribution	Formulae (Note 2)
μ	Normal	Known	Large	Normal (z)	$z = \dfrac{\bar{x} - \mu}{\left(\dfrac{\sigma}{\sqrt{n}}\right)}$
			Small		
	Unknown	Known	Large	Normal (z) via central limit theorem	
			Small	—	—
	Normal	Unknown	Small	t-distribution	$t = \dfrac{\bar{x} - \mu}{\left(\dfrac{s}{\sqrt{n}}\right)}$
			Large	Normal (z) via central limit theorem	$z = \dfrac{\bar{x} - \mu}{\left(\dfrac{s}{\sqrt{n}}\right)}$
	Unknown	Unknown	Large	Normal (z) via central limit theorem	
			Small	—	—
π	Normal	—	Large	Normal (z)	$z = \dfrac{p - \pi}{\sqrt{\dfrac{\pi(1-\pi)}{n}}}$
			Small		
	Unknown	—	Large	Normal (z) if both $n\pi$ and $n(1-\pi) \geq 5$	
			Small	χ^2-distribution	Chapter 9

Note 1:
Large sample: $n \geq 30$ Small sample: $n < 30$

Note 2:
For finite populations, multiply standard error of infinite populations formulae by finite correction factor: $\sqrt{\dfrac{N-n}{N-1}}$

(5) Stating the decision rule and defining the region of rejection based on the level of significance.

A decision rule is a statement that indicates the action to be taken, that is, accept H_0 or reject H_0 for all possible values of the test statistic.

The decision rule divides the area under the curve of the sampling distribution of the test statistic into two regions — acceptance region and rejection region — by establishing one or two critical values on the horizontal axis of the distribution. The cut-off point (*s*) between the acceptance and rejection regions is called the critical value(s) and is expressed in the same units of measurement as the test statistic (i.e. *z* or *t*). The critical value depends on the size of the rejection region which corresponds with the level of significance (α). Considering a two-tail test, the region of rejection is represented by the areas in the two tails of the curve containing $\alpha/2$ % of the sample distribution means farthest from the hypothesized mean (μ_0) — two critical values are needed, one to the left and one to the right. For the one-tail alternative hypothesis, the rejection region is placed on one side of the curve only with one critical value.

Two-tail test: normal distribution

Divide α by 2: (5 % ÷ 2 = 2.5 %) in each tail. Look up an area of 0.475 in the normal table to find the critical values of +1.96 and –1.96.

Two-tail test of H_0
$\alpha = 0.05$

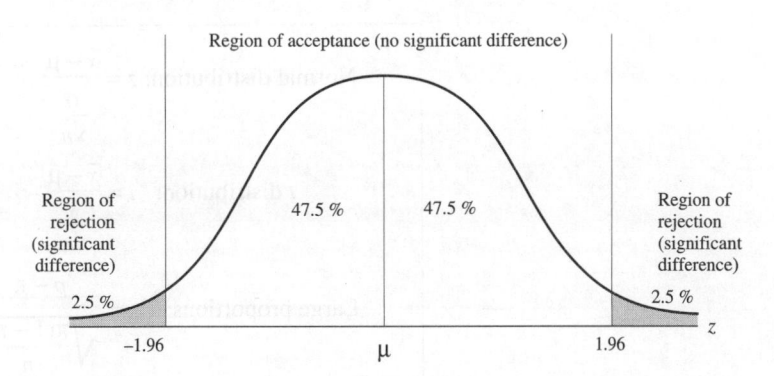

Region of acceptance (no significant difference)

Region of rejection (significant difference) 47.5 %   47.5 %   Region of rejection (significant difference)

2.5 %   2.5 %

–1.96 μ 1.96   z

$H_0: \mu = \mu_0$
$H_a: \mu \neq \mu_0$
$\alpha = 0.05$
☐ = rejection region
Reject H_0 if: $z \leq -1.96$ or $z \geq 1.96$

One-tail test

One-tail test of H₀
α = 0.05

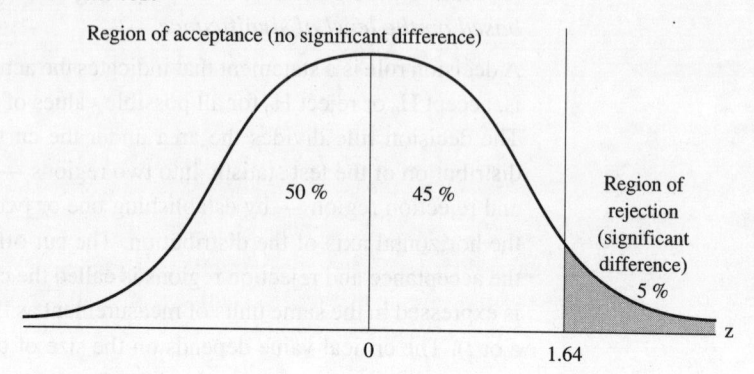

Region of acceptance (no significant difference)

50 %   45 %

Region of rejection (significant difference) 5 %

0   1.64   z

H₀: µ = µ₀
Hₐ: µ ≥ µ₀
α = 0.05
▓ = rejection region
Reject H₀ if: z test ≥ 1.64

(6) Compute the value of the test statistic.

Remember to make use of the finite population factor whenever the population is finite.

Formulae for the different test statistics:

$$\text{Normal distribution: } z = \frac{\bar{x} - \mu}{\frac{\sigma}{\sqrt{n}}}$$

$$t \text{ distribution: } t = \frac{\bar{x} - \mu}{\frac{s}{\sqrt{n}}}$$

$$\text{Large proportions: } z = \frac{p - \pi}{\sqrt{\frac{\pi(1 - \pi)}{n}}}$$

(7) Decision: accept or reject the null hypothesis by comparing the obtained value of the test statistic with the critical value (s).

By failing to reject the H₀, the investigator takes the position that there is no statistical evidence to reject it. Nevertheless, it has not been proven that the null hypothesis is true.

HYPOTHESIS TESTING

(8) State your conclusion.

Examples

Example 1 **Two-tail test**

The owner of the Kit Cake Company stated that the average number of buns sold daily was 1 500. An employee wants to test the accuracy of the boss's statement. A random sample of 36 days showed that the average daily sales were 1450 buns. Using a 1 % level of significance and assuming that $\sigma = 120$ buns, what should the employee conclude?

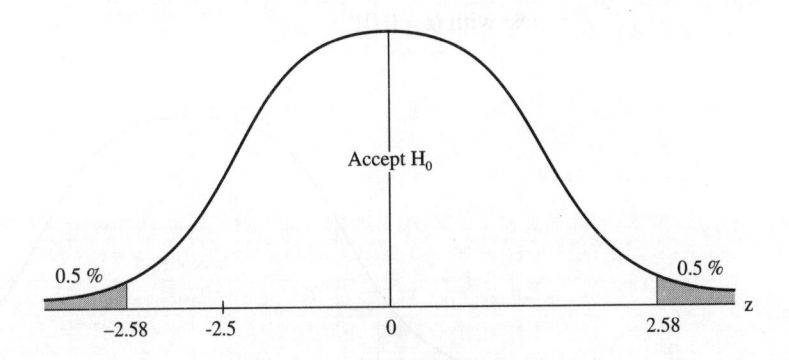

$H_0: \mu = 1\ 500$
$H_a: \mu \neq 1\ 500$
$\alpha = 0.01$

Two-tail test

The z-distribution applies because σ is known and $n > 30$.

Decision rule:

Reject H_0 if $z \leq -2.58$ or $z \geq 2.58$

Test statistic

$$z = \frac{\bar{x} - \mu}{\frac{\sigma}{\sqrt{n}}}$$

$$\frac{1\ 450 - 1\ 500}{\frac{120}{\sqrt{36}}}$$

$$= -2.5$$

Decision: Because the test statistic –2.5 is not in the rejection region, the null hypothesis should not be rejected at $\alpha = 0.01$.

Conclusion: Testing at $\alpha = 0.01$, there is no evidence to suggest that the average daily sales are different from 1 500 buns.

Example 2 One-tail test

Mr Hall, the supervisor of the local brewery, wants to make sure that the average volume of the Super can is 500 ml. If the average volume is significantly less than that, customers would be likely to complain, prompting undesirable publicity. A random sample of 36 cans showed a sample mean of 490 ml. Assuming $\sigma = 6$ ml, conduct a hypothesis test with $\alpha = 0.01$.

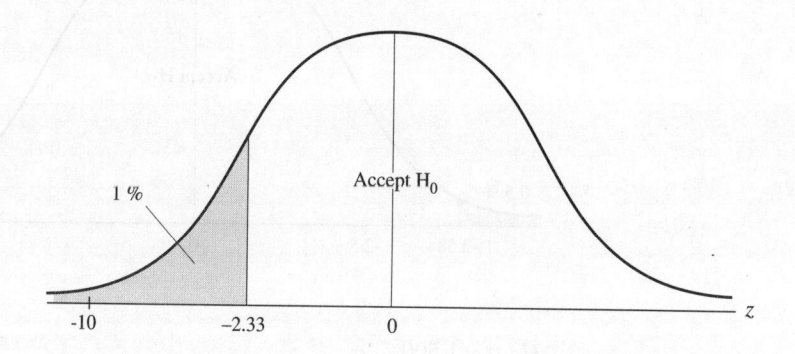

H_0: $\mu = 500$ ml
H_a: $\mu < 500$ ml
$\alpha = 0.01$
z-distribution, one-tail test on left side.

Decision rule

Reject H_0 if $z < -2.33$

Test statistic

$$z = \frac{\bar{x} - \mu}{\frac{\sigma}{\sqrt{n}}}$$

$$= \frac{490 - 500}{\frac{6}{\sqrt{36}}}$$

$$= -10$$

HYPOTHESIS TESTING

Decision: Because the test statistic −10 is in the rejection area, the null hypothesis should be rejected at α = 0.01

Conclusion: Testing at α = 0.01 there is evidence to suggest that the average volume of the Super can is significantly lower than 500 ml.

Example 3 Mr Soda thinks that his business sells an average of 17 cans of Soda Delight daily. His partner, Mr Drinkwater, thinks this estimate is wrong. A random sample of 36 days showed a mean of 15 cans and a sample standard deviation of 4 cans. Test the accuracy of Mr Drinkwater's statement at the 10 % level of significance.

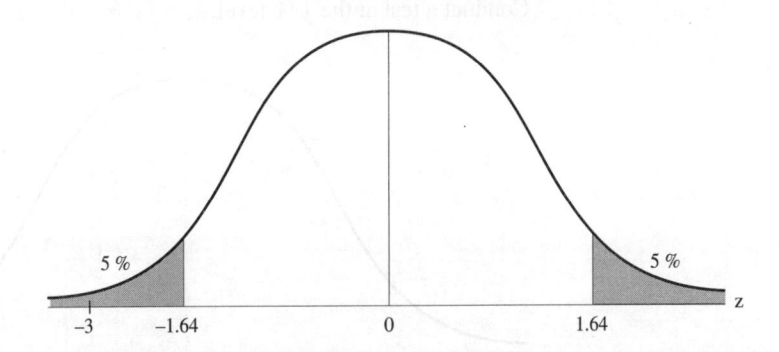

H_0: μ = 17
H_a: μ ≠ 17
α = 0.10
z-distribution; two-tail test

Decision rule

Reject H_0 if: z ≤ −1.64 or z ≥ 1.64

Test statistic

$$z = \frac{\bar{x} - \mu}{\frac{s}{\sqrt{n}}}$$

$$= \frac{15 - 17}{\frac{4}{\sqrt{36}}}$$

$$= -3$$

Decision: Because the test statistic −3 is in the rejection area, the null hypothesis should be rejected at α = 0.10.

Conclusion: Testing at $\alpha = 0.10$ there is evidence to suggest that Mr Soda's estimate of 17 cans is wrong.

Example 4   Mr Brian of an employment agency believes that the agency receives an average of 16 complaints per week from companies that employed staff through the agency. Mr Dell, an interviewer, is concerned that the true mean might be even higher. If Mr Brian's hypothesis is an understatement, something must be done about the agency's screening procedures. A sample of 10 weeks yielded an average of 18 complaints with a standard deviation of 3 complaints. Assume that the number of complaints follows the normal distribution.
Conduct a test at the 1 % level.

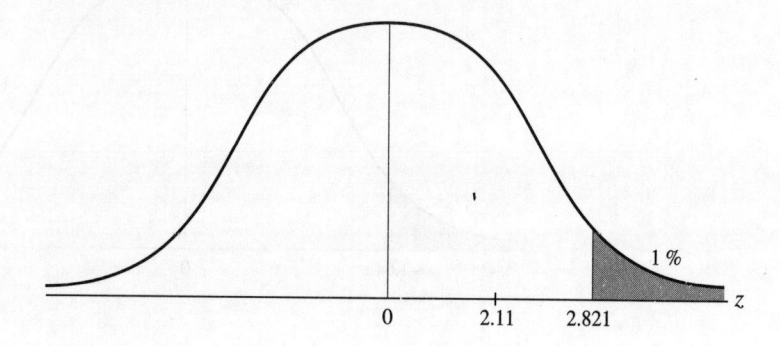

H_0: $\mu = 16$
H_a: $\mu > 16$
  $\alpha = 0.01$
t-distribution: one-tail test with 9 degrees of freedom

Decision rule
Reject H_0 if $t > 2.821$

Test statistic

$$t = \frac{\bar{x} - \mu}{\frac{s}{\sqrt{n}}}$$

$$= \frac{18 - 16}{\frac{3}{\sqrt{10}}}$$

$$= 2.11$$

Decision: Because the test statistic 2.11 is not in the rejection area, the null hypothesis should not be rejected at $\alpha = 0.01$.

Conclusion: Testing at α = 0.01 there is no evidence to suggest that the average number of complaints per week is different from 16.

Example 5 *Hypothesis test of proportions:*
A local newspaper has stated that only 25 % of all college students read newspapers daily. A random sample of 200 students showed that 45 of them were daily readers of newspapers. Test the accuracy of the newspaper's statement using a significance level of 5 %.

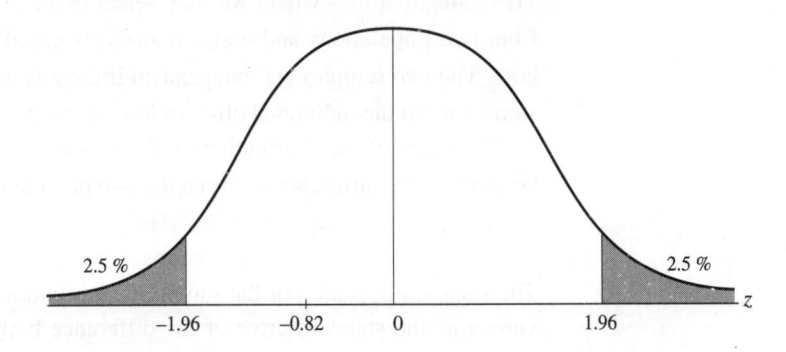

[Normal distribution curve with 2.5% shaded in each tail at −1.96 and 1.96; −0.82 marked on horizontal axis]

H_0: π = 0.25

H_a: π ≠ 0.25

α = 0.05

z-distribution; two-tail test of proportions

Decision rule

Reject H_0 if z ≤ −1.96 or z ≥ 1.96

Test statistic

$$z = \frac{p - \pi}{\sqrt{\frac{\pi(1-\pi)}{n}}}$$

$$= \frac{0.225 - 0.25}{\sqrt{\frac{0.25 \times 0.75}{200}}}$$

$$= -0.82$$

Decision: Because the test statistic −0.82 is not in the rejection area, the null hypothesis should not be rejected at α = 0.05.

Conclusion: Testing at $\alpha = 0.05$ there is no evidence to suggest that the proportion of daily readers of newspapers is different from 25 %.

8.3 INFERENCES INVOLVING TWO POPULATIONS

There are situations where we may select two independent samples from two populations and perform analyses based on the results of both. The two samples are independent if there is no known relationship between the individual observations of the two samples.

The mean of the distribution of differences in sample means will be equal to the difference between the two population means:

$$\mu_{\bar{x}_1 - \bar{x}_2} = \mu_1 - \mu_2$$

The standard deviation of the sampling distribution of differences is known as the standard error of the difference between two sample means

$$\sigma_{\bar{x}_1 - \bar{x}_2} = \sqrt{\frac{\sigma_1^2}{n_1} + \frac{\sigma_2^2}{n_2}}$$

In hypothesis tests with two samples, the hypothesis procedure is the same as in tests with one sample. However, the H_0, H_a and formulae are slightly different.

We are testing the null hypothesis (H_0) against the following alternative hypothesis (H_a)

$H_0: \mu_1 = \mu_2$ $\quad\quad$ $H_0: \pi_1 = \pi_2$
$H_a: \mu_1 \neq \mu_2$ $\quad\quad$ $H_a: \pi_1 \neq \pi_2$ or
$H_a: \mu_1 \geq \mu_2$ $\quad\quad$ $H_a: \pi_1 \geq \pi_2$ or
$H_a: \mu_1 \leq \mu_2$ $\quad\quad$ $H_a: \pi_1 \leq \pi_2$

❏ If:

(1) both populations are normal, or

(2) both populations are normal via the central limit theorem, and

(3) the standard deviations of the two populations are known:

the sampling distribution of the differences will follow the normal distribution.

$(1 - a)$ % confidence interval estimate for $\mu_1 - \mu_2$:

$$\mu_1 - \mu_2 = \bar{x}_1 - \bar{x}_2 \pm z \cdot \sqrt{\frac{\sigma_1^2}{n_1} + \frac{\sigma_2^2}{n_2}}$$

Test statistic for H_0: $\mu_1 = \mu_2$:

$$z = \frac{\bar{x}_1 - \bar{x}_2}{\sqrt{\frac{\sigma_1^2}{n_1} + \frac{\sigma_2^2}{n_2}}}$$

❏ If:
(1) the standard deviations of the two populations are unknown,
(2) the shapes of the population distributions are not necessarily normal, and
(3) both sample sizes are large (n_1 and $n_2 \geq 30$) so that the central limit theorem applies,

the sampling distribution of the differences will follow the normal distribution.

$(1 - a)$ % confidence interval estimate for $\mu_1 - \mu_2$:

$$\mu_1 - \mu_2 = \bar{x}_1 - \bar{x}_2 \pm z \cdot \sqrt{\frac{s_1^2}{n_1} + \frac{s_2^2}{n_2}}$$

Test statistic for H_0: $\mu_1 = \mu_2$:

$$z = \frac{\bar{x}_1 - \bar{x}_2}{\sqrt{\frac{s_1^2}{n_1} + \frac{s_2^2}{n_2}}}$$

❏ If:
(1) the standard deviations of the two populations are unknown but equal ($\sigma_1^2 = \sigma_2^2$),
(2) the shapes of the population distributions are normal, and
(3) both sample sizes are small (n_1 and $n_2 < 30$),

the sampling distribution of the differences will follow the *t*-distribution.

$(1 - a)$ % confidence interval estimate for $\mu_1 - \mu_2$:

$$\mu_1 - \mu_2 = \bar{x}_1 - \bar{x}_2 \pm t \cdot \sqrt{\frac{(n_1 - 1)s_1^2 + (n_2 - 1)s_2^2}{n_1 + n_2 - 2}} \cdot \sqrt{\frac{1}{n_1} + \frac{1}{n_2}}$$

Test statistic for H_0: $\mu_1 = \mu_2$:

$$t = \frac{\bar{x}_1 - \bar{x}_2}{\sqrt{\frac{(n_1 - 1)s_1^2 + (n_2 - 1)s_2^2}{n_1 + n_2 - 2}} \cdot \sqrt{\frac{1}{n_1} + \frac{1}{n_2}}}$$

Note:

- A pooled estimate of the standard deviation is used, which means that the information from both samples is combined so as to give the best possible estimate.
- The number of degrees of freedom = $n_1 + n_2 - 2$
- The two population variances (σ_1^2 and σ_2^2) are unknown, therefore we will employ the F-test for homogeneity of variances in chapter 10 to determine whether the variances are equal or not.

- If:

  (1) the standard deviations of the two populations are unknown but not equal ($\sigma_1^2 \neq \sigma_2^2$),
  (2) the shapes of the population distributions are normal, and
  (3) both sample sizes are small (n_1 and $n_2 < 30$),
  the sampling distribution of the differences will follow the *t*-distribution.

  $(1 - a)$ % confidence interval estimate for $\mu_1 - \mu_2$:

  $$\mu_1 - \mu_2 = \bar{x}_1 - \bar{x}_2 \pm t \cdot \sqrt{\frac{s_1^2}{n_1} + \frac{s_2^2}{n_2}}$$

HYPOTHESIS TESTING

Test statistic for $H_0: \mu_1 = \mu_2$:

$$z = \frac{\bar{x}_1 - \bar{x}_2}{\sqrt{\dfrac{s_1^2}{n_1} + \dfrac{s_2^2}{n_2}}}$$

Note:

The number of degrees of freedom for the critical value is given by the smaller of $n_1 - 1$ or $n_2 - 1$.

❏ If we are interested in making statistical comparisons between the proportions or percentages associated with two populations and n_1 and n_2 are both large samples ($n_1p_1, n_1q_1, n_2p_2, n_2q_2$ should all be ≥ 5), the sampling distribution of the difference between two sample proportions $(p_1 - p_2)$ is normally distributed.

$(1 - a)$ % confidence interval estimate for $\pi_1 - \pi_2$:

$$\pi_1 - \pi_2 = p_1 - p_2 \pm z \cdot \sqrt{\dfrac{p_1 q_1}{n_1} + \dfrac{p_2 q_2}{n_2}}$$

Note:

π_1 and π_2 are unknown and p_1 and p_2 are used in the equation to estimate the standard error of the difference between two sample proportions.

Test statistic for $H_0: \pi_1 = \pi_2$:

$$z = \frac{\pi_1 - \pi_2}{\sqrt{\hat{p} \cdot \hat{q} \cdot \left(\dfrac{1}{n_1} + \dfrac{1}{n_2}\right)}}$$

$$\text{where } \hat{p} = \frac{n_1 p_1 + n_2 p_2}{n_1 + n_2}$$

$$\text{and } \hat{q} = 1 - \hat{p}$$

Note:

Since p_1 and p_2 are unknown but assumed equal in the H_0, the best estimate for the standard error of the difference between two sample proportions is obtained by pooling the two samples.

Examples

Example 1   Darlow Stores owns outlet A and outlet B. For the past year, outlet A has spent more rands on advertising widgets than outlet B. The corporation wants to determine whether the advertising has resulted in more sales for outlet A. A sample of 36 days at outlet A had a mean of 170 widgets sold daily. A sample of 36 days at outlet B had a mean of 165. Assuming $\sigma_1 = 6$ and $\sigma_2 = 5$, what would be concluded if a test were conducted at the 5 % level of significance?

$H_0: \mu_1 = \mu_2$

$H_a: \mu_1 > \mu_2$

$\alpha = 0.05$

z-distribution: one-tail test on the right side

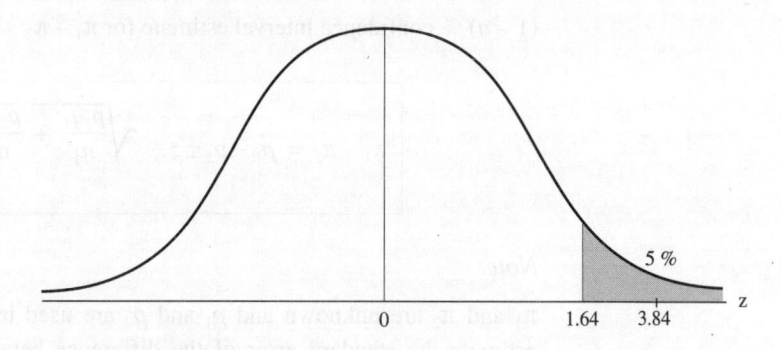

Decision rule

Reject H_0 if $z > 1.64$

Test statistic

$$z = \frac{\bar{x}_1 - \bar{x}_2}{\sqrt{\dfrac{\sigma_1^2}{n_1} + \dfrac{\sigma_2^2}{n_2}}}$$

HYPOTHESIS TESTING

$$= \frac{170 - 165}{\sqrt{\frac{6^2}{36} + \frac{5^2}{36}}}$$

$$= 3.84$$

Decision: Because the test statistic 3.84 is in the rejection region, the null hypothesis should be rejected at $\alpha = 0.05$.

Conclusion: Testing at $\alpha = 0.05$ there is evidence to believe that outlet A has sold more widgets than outlet B.

Construct a 95 % confidence interval estimate for the difference between the mean for outlet A and the mean for outlet B.

$$\mu_A - \mu_B = \bar{x}_A - \bar{x}_B \pm z \cdot \sqrt{\frac{\sigma_A^2}{n_A} + \frac{\sigma_B^2}{n_B}}$$

$$= 170 - 165 \pm 1.96 \cdot \sqrt{\frac{6^2}{36} + \frac{5^2}{36}}$$

$$= 5 \pm 2.55$$

$$2.45 \le \mu_A - \mu_B \le 7.55$$

With 95 % confidence we estimate the difference between the two means to be between −2.45 and +7.55.

Example 2 A manufacturer of two types of shoes (A and B) is interested to test the hypothesis that the average price of type A is less than the average price of type B. A random sample of 12 retailers who stock type A yielded an average price of R26 with a standard deviation of R2. A random sample of 10 retailers who stock type B yielded a mean price of R28 with a standard deviation of R4. Test the hypothesis at a 5 % level of significance. Assume the two samples came from two normally distributed populations with equal variances.

$H_0: \mu_1 = \mu_2$
$H_a: \mu_1 < \mu_2$
    $\alpha = 0.05$

t-distribution: one-tail test with 20 degrees of freedom

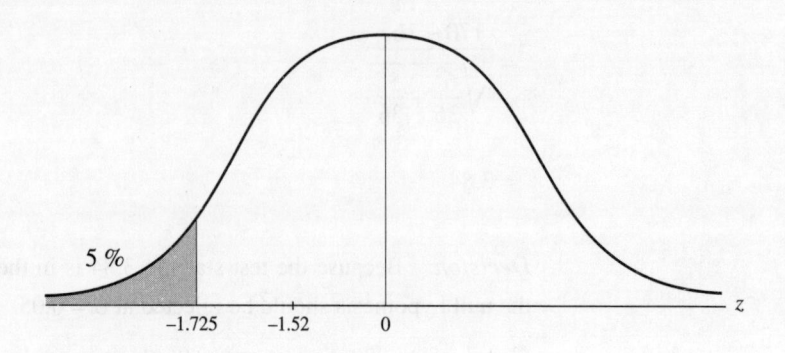

Decision rule

Reject H_0 if $t < -1.725$

Test statistic

$$t = \frac{\bar{x}_1 - \bar{x}_2}{\sqrt{\frac{s_1^2 (n_1 - 1) + s_2^2 (n_2 - 1)}{n_1 + n_2 - 2}} \cdot \sqrt{\frac{1}{n_1} + \frac{1}{n_2}}}$$

$$= \frac{26 - 28}{\sqrt{\frac{2^2 (12 - 1) + 4^2 (10 - 1)}{12 + 10 - 2}} \cdot \sqrt{\frac{1}{12} + \frac{1}{10}}}$$

$$= -1.52$$

Decision: Because the test statistic -1.52 is not in the rejection region, the null hypothesis should not be rejected at $\alpha = 0.05$.

Conclusion: Testing at $\alpha = 0.05$ there is no evidence to suggest that type A is cheaper than type B.

Construct a 95 % confidence interval estimate for the difference between the mean price for type A and the mean price for type B.

$$\mu_A - \mu_B = \bar{x}_A - \bar{x}_B \pm t \cdot \sqrt{\frac{(n_A - 1)s_A^2 + (n_B - 1)s_B^2}{n_A + n_B - 2}} \cdot \sqrt{\frac{1}{n_A} + \frac{1}{n_B}}$$

$$= 26 - 28 \pm 2.086 \cdot \sqrt{\frac{(12 - 1)2^2 + (10 - 1)4^2}{12 + 10 - 2}} \cdot \sqrt{\frac{1}{12} + \frac{1}{10}}$$

$$= -2 \pm 2.74$$

$$-4.74 \leq \mu_A - \mu_B \leq 0.74$$

HYPOTHESIS TESTING

With 95 % confidence we estimate the difference between the two means to be between −4.74 and +0.74.

Example 3 The manufacturer of Munchy believes that her product will be more popular in region Y than in region X, where it is currently being produced and distributed. In order to check this hypothesis, a random sample was taken from each region. The sample in region X contained 700 people, 560 of whom claimed to prefer the taste of Munchy. In region Y, 525 of the 750 people sampled, responded favourably. Based on these results, does it appear that Munchy will be more popular in region Y than in region X?

$p_x = 0.8$

$p_y = 0.7$

$$\hat{p} = \frac{p_y(n_y) + p_x(n_x)}{n_y + n_x}$$

$$= \frac{0.7(750) + 0.8(700)}{1\ 450}$$

$$= 0.75$$

$H_0: \pi_y = \pi_x$

$H_a: \pi_y > \pi_x$

$\alpha = 0.05$

z-distribution: one-tail test of proportion

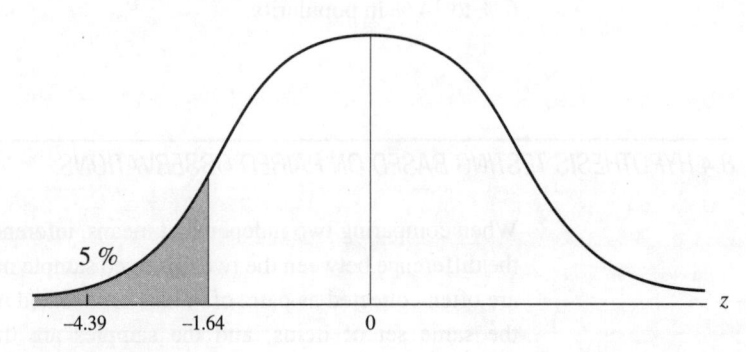

5 %

−4.39   −1.64   0   z

Decision rule

Reject H_0 if $z > 1.64$

Test statistic

$$z = \frac{p_y - p_x}{\sqrt{\hat{p}\hat{q}\left(\frac{1}{n_y} + \frac{1}{n_x}\right)}}$$

$$= \frac{0.7 - 0.8}{\sqrt{(0.75)(0.25)\left(\frac{1}{750} + \frac{1}{700}\right)}}$$

$$= -4.39$$

Decision: Because the test statistic −4.39 is in the acceptance region, the null hypothesis should be accepted at $\alpha = 0.05$.

Conclusion: Testing at $\alpha = 0.05$ there is no evidence to suggest that the product is doing better in the new market.
Construct a 95 % confidence interval estimate for the difference between the popularity in region Y and the popularity in region X

$$\pi_y - \pi_x = p_y - p_x \pm z \cdot \sqrt{\frac{p_y q_y}{n_y} + \frac{p_x q_x}{n_x}}$$

$$= 0.7 - 0.8 \pm 1.96 \cdot \sqrt{\frac{0.7(0.3)}{750} + \frac{0.8(0.2)}{700}}$$

$$= -0.1 \pm 0.04$$
$$= 0.14 \leq \pi_y - \pi_x \leq -0.06$$

With 95 % confidence we estimate that there is a difference of from 6 % to 14 % in popularity.

8.4 HYPOTHESIS TESTING BASED ON PAIRED OBSERVATIONS

When comparing two independent means, inferences are made about the difference between the two observed sample means. But, samples are often collected as pairs of values or repeated measurements from the same set of items, and the samples are therefore dependent samples.

In either case, the variable of interest becomes the difference in each pair of observations rather than the value of the individual observations. This difference is called a paired difference.

The distribution of this population of differences is assumed to be normal with an unknown standard deviation.

Before conducting the hypothesis test, determine the difference (d) between each pair of values in the paired samples, and square the differences.

HYPOTHESIS TESTING

Example 1 Advertisements for the Rapid Fitness Centre claim that completion of the centre's course will result in weight loss. A random sample of 8 recent members revealed the following body weights before and after completion of the course. At a 0.01 significance level, can it be concluded that the course will result in a significant weight loss?

Name	Before (kg)	After (kg)	d (Before – After)	d^2
Will	75	74	1	1
Sue	110	100	10	100
Lou	70	73	–3	9
Mike	80	78	2	4
Ron	105	97	8	64
Bruce	91	88	3	9
Lynn	81	74	7	49
Con	85	81	4	16
			32	252

$H_0: \mu_d = 0$ (μ_d = assumed value of population difference)
$H_a: \mu_d > 0$
$\alpha = 0.01$

Decision rule

Reject H_0 if $t > 2.998$ with $n = 8$ and $df = 7$.
Note: n is the number of paired differences and $df = n - 1$.

Test statistic

$$t = \frac{\bar{d} - \mu_d}{\frac{s_d}{\sqrt{n}}} \quad \text{with} \quad \bar{d} = \frac{\sum d}{n}$$

$$= \frac{32}{8}$$

$$= 4$$

$$s_d = \sqrt{\frac{n\left(\sum d^2\right) - \left(\sum d\right)^2}{n(n-1)}}$$

$$= \sqrt{\frac{8(252) - (32)^2}{8(7)}}$$

$$t = \frac{4-0}{\left(\frac{4.21}{\sqrt{8}}\right)} = 4.21$$

$$= 2.69$$

Decision: Because the test statistic 2.69 is not in the rejection region, the null hypothesis should not be rejected at $\alpha = 0.01$.

Conclusion: The course cannot be shown to result in a significant weight loss.

Example 2 A tyre company wishes to determine whether there is evidence that the tread life is lower for tyres driven at an average speed of 130 km/h than for tyres driven at an average speed of 110 km/h. A pair of tyres from each of eight different types is selected. One tyre of each type is driven at 130 km/h, and the other at 110 km/h. The results are shown below.

Type	Tyre life ('000 km) 130 km/h	110 km/h	d	d^2
1	24.31	26,42	−2.11	4.45
2	31.27	33.77	−2.50	6.25
3	30.71	35.42	−4.71	22.18
4	28.64	30.32	−1.68	2.82
5	23.60	22.85	0.75	0.56
6	36.41	42.71	−6.3	39.69
7	21.46	25.09	−3.63	13.18
8	30.62	31.76	−1.14	1.30
			−21.32	90.43

$H_0: \mu_d = 0$
$H_a: \mu_d < 0$ (i.e. $\mu_{130} < \mu_{110}$)
$\alpha = 0.01$

Decision rule

Reject H_0 if test statistic < -2.998 with $n = 8$ and $df = 7$.

Test statistic

$$t = \frac{\bar{d} - \mu_d}{\left(\frac{s_d}{\sqrt{n}}\right)} \quad \text{with} \quad \bar{d} = \frac{\sum d}{n}$$

$$= \frac{-21.32}{8}$$

$$= -2.665$$

$$s_d = \sqrt{\frac{n\left(\sum d^2\right) - \left(\sum d\right)^2}{n(n-1)}}$$

$$= \sqrt{\frac{8(90.43) - (-21.32)^2}{8(7)}}$$

$$= 2.19$$

$$= \frac{-2.665 - 0}{\left(\frac{2.19}{\sqrt{8}}\right)}$$

$$= -3.44$$

Decision: Because the test statistic -3.44 is in the rejection region, the null hypothesis should be rejected at $\alpha = 0.01$.

Conclusion: There is evidence to suggest that the tyre life is lower when driven at 130 km/h than when driven at 110 km/h.

EXERCISES

Exercise 1 A new cough medicine is being tested. A sample of 50 people yields an average recovery time of 2.86 days with a standard deviation of 0.8 days. Are these results consistent with the company's claim that the average recovery time is 2.8 days? Assume a 2 % level of significance.

Exercise 2 During the past several years, frequent checks were made of the spending patterns of tourists returning from vacations to countries in Asia. Results indicated that travellers spent an average of R1 010 on items such as meals, films, gifts, and so on. A new survey is to be conducted to determine whether there has been a change in the average amount spent. A sample of 50 travellers has a sample mean of R1 090 and a standard deviation of R300. At the 0.01 level, is there evidence that there has been a change in the mean amount spent abroad, or is the increase of R80 due to chance?

Exercise 3   Records show that the average mark of students taking statistics is 65 % with a standard deviation of 17 %. A new method of teaching was developed. A sample of 64 students was selected and the sample mean was 62 %. Is the new teaching method at least as successful as the old method at a significance level of 0.05?

Exercise 4   Hyperactive children are often disruptive in the typical classroom setting because they find it difficult to remain seated for extended periods of time. Baseline data from a very large study show that the typical frequency of 'out-of-seat behaviours' was 12.38 per hour with a standard deviation of 3.52. Treatment was applied to a group of 30 hyperactive children. The average number of 'out-of-seat behaviours' was reduced to 11.59 per hour. Using $\alpha = 1$ %, can we conclude that this decline in 'out-of-seat behaviours' is significant?

Exercise 5   A machine is set to fire 30 g of dried fruit into a box of cereal moving along the production line. Of course, there is some variation in the mass of the fruit inserted. A sample of 36 boxes of mix revealed that the average mass of the fruit inserted was 30.2 g with a sample standard deviation of 0.50 g. Is the increase in the mass of the fruit inserted significant at the 0.05 level?

Exercise 6   The school committee of a country school wants to consider a new sports programme funded by the Department of Education. In order to be eligible for the government grant, the arithmetic mean income per household should not exceed R16 000 per year. In a survey including 75 households the mean income was R17 000 and the standard deviation of the sample was R3 000. Using the 0.01 level of significance, can the committee argue that the larger household income is due to chance?

Exercise 7   Mag Wheels advertises extensively that they can fit and balance tyres on any car in 30 min or less on the average. An Advertising Watchdog Group investigates this claim. A fleet of ten cars, all unmarked and in need of tyres, are dispatched to nearby Mag Wheel shops. The average time taken on each of the ten cars is determined as 31 min with a standard deviation of 4.24 min. Does this evidence suggest that Mag Wheels is unable to meet its advertised claim? ($\alpha = 0.05$)

HYPOTHESIS TESTING

Exercise 8   Club 60 said that typical senior citizens living at high altitudes claim that their systolic blood pressure is lower than the average of 160. To test this claim, 20 senior citizens living at high altitudes were selected at random and their blood pressure checked. Their mean systolic pressure was 151, and the standard deviation 12. Do senior citizens living at high altitudes have significantly lower systolic blood pressure? Use the 0.05 level.

Exercise 9   A car manufacturer asserts that with the new rubber bumper system, the average body repair cost for the damage sustained in a collision at 15 km/h does not exceed R400. To test the validity of this claim, six cars are crashed into a barrier at 15 km/h and their repair costs recorded. The mean and the standard deviation are found to be R458 and R48, respectively. At the 0.05 level of significance, do the test data contradict the manufacturer's claim that the repair cost does not exceed R400?

Exercise 10   A new toy has been developed and the manufacturer hopes to market it for the coming Christmas season. Before going into full production, a large number of toys were sent to five test market areas. The manufacturer plans to start full production if the monthly sales in the test markets average more than R20 000 during the one month trial period. The results were R20 000; R16 000; R25 000; R19 000; R24 000. Using the 0.05 level of significance, test whether the manufacturer should start with full production.

Exercise 11   It is claimed that 90 % of people would continue to work even if they inherited sufficient wealth to live comfortably without working. This assertion is questioned by a social research team. A list of persons who recently inherited more than R400 000 is obtained, and 350 persons are selected from the list. Out of the 350 selected, 308 indicate that they are still working. At the 0.05 level, is this sufficient evidence to doubt your claim?

Exercise 12   The advertising director of a TV station claims that at least 50 % of viewers will remember the name of a product advertised on his station. A sample of 100 viewers were taken and 45 of them recalled the name of a certain product that was advertised. Test at a 2 % level of significance if the claim is correct.

Exercise 13   A local newspaper conducted a study of unemployment claims. It reported that at least 20 % of current recipients of unemployment benefits are not eligible to receive such benefits. To investigate the newspaper's claim, you randomly select 400 recipients from the files and carefully investigate each one. You find that 60 should not be receiving payments because they either failed to report that they work, or committed other fraudulent acts. At the 0.01 level of significance, should the newspaper's claim be rejected?

Exercise 14   Your boss complains that you make more typing errors than the normal 2 %. A sample of 30 pages indicates 7 pages with errors. Is the boss's claim justified at a 10 % significance level?

Exercise 15   It is known that approximately 1 in 10 smokers favours cigarette brand A. After a promotional campaign in a given sales region, a sample of 200 cigarette smokers was interviewed to determine the effectiveness of the campaign. The result of the survey showed that a total of 26 people expressed a preference for brand A. Do these data present sufficient evidence to indicate an increase in the acceptance of brand A in the region? ($\alpha = 0.05$)

Exercise 16   A manufacturer has found defective items in recent production lots. He allows for up to 5 % defectives. In a sample of 36 lots he finds 6 % defective. Test at a level of 0.10 whether the number of defectives exceeds the allowed tolerance.

Exercise 17   Mr Admission wants to investigate whether there is any difference between the mathematics marks of students who attend lectures during the day and the marks of students who attend evening classes.

Student	Mean marks	Standard deviation	Sample size
Day	90	12	40
Evening	94	15	50

Use the 0.05 level of significance.

Exercise 18   Elize is interested in knowing whether typists required to correct their own errors have the same error rate as typists not required to correct their own errors. In a sample of 40 typists required to correct their own errors, the average number of errors per day is determined as 20.2 with a standard deviation of 2.5. The sample of 56 typists not required to correct their own errors yielded an average of 21.0 errors per day with a standard deviation of 3.1. At a significance level of 10 %, is there a noteworthy increase in the number of errors made by the second group?

Exercise 19   Two machines fill bottles with a cough syrup. The performance of each is checked periodically by testing bottles removed at random from the production line. A sample of 40 bottles for machine 1 showed a mean mass of 202.6 ml and standard deviation of 3.3 ml. For machine 2, a sample of 50 bottles showed a mean mass of 200.0 ml and standard deviation of 2.0 ml. Using the 0.01 level of significance, test whether there is any difference in the performance of the two machines.

Exercise 20   A medical researcher claims that lung capacity varies significantly between smokers and non-smokers. The mean capacity of a sample of 30 non-smokers is 5.0 ℓ with a sample standard deviation of 0.3 ℓ. Of the 40 smokers in the sample, the mean lung capacity is 4.5 ℓ with a standard deviation of 0.4 ℓ. At the 0.01 significance level, is there sufficient evidence to conclude that lung capacity is larger among non-smokers?

Exercise 21   An assembly operation in a factory requires training period for new employees of approximately 3 months. A new method of training was suggested and a test was conducted to compare the new method with the standard procedure. Two groups of nine new employees were trained. One group used the new method and the other used the standard training procedure. The length of time in minutes required for each employee to assemble the device was recorded at the end of the training period. These measurements appear in the following table. Do the data present sufficient evidence to indicate that the mean assembly is reduced by the new method? ($\alpha = 0.05$)

Standard method: 32 37 35 28 41 44 35 31 34
New method: 35 31 29 25 34 40 27 32 31

Exercise 22   An experiment was conducted to compare the mean lengths of time required by two employees to open new personal cheque accounts for clients. Ten clients were randomly assigned to each employee and the time taken to serve each client was recorded in minutes. The means and variances for the two samples were as follows:

	Employee A	Employee B
Mean:	22.2	28.5
Variance:	16.36	18.92

Do the data provide sufficient evidence to indicate a difference in mean times required to complete the paperwork necessary for a new cheque account? Test using $\alpha = 0.10$

Exercise 23   Two methods of memorizing difficult material are being tested. Nine pairs of students, randomly selected, are each assigned to one of the two methods. A test is finally given to all the students, with the following results:

Method A: 90   86   72   65   44   52   46   38   43
Method B: 85   87   70   62   44   53   42   35   46

Using a 5 % significance level, test to determine if there is a difference in the effectiveness of the two methods.

Exercise 24   Measurements of the left-handed and right-handed gripping strength of ten left-handed people are recorded.

Left hand: 140   90   125   130   95   121   85   97   131   110
Right hand: 138   87   110   132   96   120   86   90   129   100

Do these data provide evidence, at the 0.01 level of significance, that those tested have greater gripping strength in their dominant hand?

Exercise 25   The effectiveness of a newly developed allergy-relief capsule is to be compared with that of one which has been on the market for a number of years. A sample of 250 persons using the new capsule revealed that 150 obtained satisfactory relief. Out of a group of 400 using the older capsule, 232 obtained satisfactory relief. Using the 2 % level of significance, test the hypothesis that the proportion obtaining relief from the new capsule is equal to the proportion obtaining relief from the old capsule.

Exercise 26   A major car insurance company claims that, compared with other drivers, a larger proportion of younger drivers take serious risks when driving. To investigate this claim, you give a driving-simulator test to a sample of young drivers and to a sample of other drivers. Out of 100 young drivers tested, 30 took serious risks. Out of 200 other drivers tested, 55 took serious risks. Do you think there is sufficient evidence to support the insurance company's contention? ($\alpha = 0.05$)

Exercise 27   Suppose that a random sample of 1 000 South African-born citizens had revealed that 200 favour full diplomatic relations with America. Similarly, 110 out of a sample of 500 foreign-born citizens favour it. Test at the 0.05 level the hypothesis that there is no significant difference between the two proportions.

Exercise 28   The Sigar Tobacco Co. advertises that their brand contains no more than 20 mg of nicotine on the average. A random sample of 49 cigarettes is obtained and analysed. The results revealed a mean nicotine content of 21.1 mg and a standard deviation of 3.6 mg. Test whether the sample results provide sufficient evidence at a 0.05 significance level to reject the advertised claim.

Exercise 29 A scientist researching giraffes has been gathering data from one very large herd. Due to some rather obvious problems in measuring the length of a giraffe's neck, he has been able to measure the neck lengths of only 25 giraffes chosen randomly from the herd. The mean neck length of the sample is determined as 2.98 m with a standard deviation of 0.6 m. At the 0.05 level of significance, is it reasonable of the scientist to conclude that the mean neck length of all the giraffes in the herd is 2.8 m?

Exercise 30 The administrator of a hospital believes the mean time spent in the casualty room by patients who are not admitted to the hospital is 3.0 hours. In a random sample of 40 patients who required emergency treatment last month, but were not admitted to the hospital, the mean was 2.75 hours with a standard deviation of 0.8 hours. Can it be concluded that the administrator over-estimated the mean time spent at the hospital by casualty patients? ($\alpha = 0.01$)

Exercise 31 The registrar at a teknikon is comparing the average marks obtained by students older than 21 with those of students under 21. The following are the results.

	Under 21	Over 21
Sample means	63	68
Sample standard deviation	4	4
Number in sample	60	50

At a 0.05 significance level, is there a difference between the average marks obtained by students over 21 and those of students under 21?

Exercise 32 A comparison has been made of the cost of renting a flat in Hillbrow as against Berea. A random sample of 35 units in Hillbrow revealed a mean rent of R660 per month with a standard deviation of R50. A random sample of 40 comparable units in Berea revealed a mean of R700 per month with a standard deviation of R40. Can it be concluded that the monthly rental cost in Berea is higher? Use the 0.05 level of significance.

Exercise 33 A soft drink filling machine is designed to fill bottles with 175 ml of liquid. Each machine is tested before it is shipped. If a random sample of 100 bottles yielded a sample mean of 180 ml with a sample standard deviation of 4 ml, is the machine properly adjusted? ($\alpha = 0.05$)

Exercise 34   The manager of a large hotel believes that the average length of stay of hotel guests is at least 5 days. If a random sample of 25 guest records indicates a sample mean of 4.6 days and a sample standard deviation of one day, do the results support the manager's claim? Set $\alpha = 0.05$.

Exercise 35   The manufacturer of a battery-operated car has made modifications to the motor of the car. One result of the modification seems to be extended battery life. From past data it is known that the battery in the car had an average operating life of 30 hours. To investigate if the modifications did improve battery life, a random sample of 100 cars with the new motor was obtained yielding a mean operating life of 31 hours and a standard deviation of 0.75 hours. At the 0.05 level of significance, can it be inferred that the battery life was extended?

Exercise 36   A chicken producer claims that the average mass of a particular group of chickens exceeds 1 kg. Before agreeing to a purchase, a particular customer selected a sample of 25 chickens, which yielded a sample mean of 1.12 kg and a standard deviation of 0.1 kg. If the masses can be considered to be normally distributed, should the claim be rejected at the 0.05 level of significance?

Exercise 37   A bottle-filling process has a setting of 200 ml. A random sample of 49 bottles produced a sample mean of 195 ml and a standard deviation of 3.5 ml. At a 5 % level of significance, what can be concluded about the machine setting?

Exercise 38   An electric motor was designed to have a trouble-free life of 2 000 hours. A random sample of 25 motors experienced malfunction at an average of 1 600 hours with a standard deviation of 400 hours. If $\alpha = 0.10$, what can be concluded about the trouble-free life?

Exercise 39   Credit department records indicate that the average balance for customers with accounts is R65. If a recent sample of 64 customers yielded a sample mean of R70 and a sample standard deviation of R6, is the average balance increasing? Set $\alpha = 0.05$.

Exercise 40   A company plans to market nationally a new product only if more than 70 % of the test market responds favourably to the product. What can be concluded at the 0.05 level of significance if a random sample of 1 000 people from the test market yielded 730 favourable responses?

HYPOTHESIS TESTING

Exercise 41 A cafe owner believes more than 70 % of the purchases exceed R5. If a random sample of 60 shoppers contained 48 people who bought for more than R5, is the owner's claim acceptable? ($\alpha = 0.02$)

Exercise 42 A company that operates a chain of restaurants is considering the purchase of the Pizza Inn chain. A random sample of 10 restaurants already in the chain showed an average volume of 1400 meals per day with a standard deviation of 120, whereas a survey of 15 of the Pizza Inn's showed a mean volume of 1675 meals with a standard deviation of 250. At the 0.05 level of significance, would the company be safe in acquiring the Pizza Inn chain if it desires comparable operations?

Exercise 43 The manager of a service station is interested in comparing activity in the morning and afternoon. A random sample of 12 mornings produced an average number of 18 customers with a standard deviation of 3, while a random sample of 14 afternoons yielded an average number of 22 customers with a standard deviation of 4. At the 0.05 level of significance, are the morning and afternoon activities different?

Exercise 44 The management of a very large factory wishes to investigate the effect of the 4-day work week on absenteeism. Two random samples of size 40 were selected: employees of group I worked 10-hour days (4-day week) and employees of group II worked 8-hour days (5-day week). If group I averaged 4 hours of absenteeism per week with a standard deviation of 1.2 and group II averaged 4.4 hours of absenteeism per week with a standard deviation of 1.5, should we conclude that the shorter work week reduces absenteeism?
Set $\alpha = 0.05$.

Exercise 45 A manager wants to know whether it will take less time to type a standard monthly report on a word processor than on a standard electric typewriter. A random sample of 7 secretaries is selected and the typing time (in hours) recorded. The results are as follows:

Secretary	A	B	C	D	E	F	G
Word processor	6.3	7.1	6.5	6.0	5.0	7.2	7.0
Electric typewriter	7.0	6.9	7.5	6.5	6.0	8.0	7.5

At a 0.05 level of significance, is there evidence for the manager to conclude that typing time on the word processor is less than on the electric typewriter?

CHAPTER 9
The Chi-square Test

CONTENTS

9.1 INTRODUCTION 159
9.2 TEST OF INDEPENDENCE 160
9.3 GOODNESS-OF-FIT TESTS 162
9.4 TESTS CONCERNING SMALL PROPORTIONS 168

9.1 INTRODUCTION

The chi-square distribution (χ^2) is a probability distribution which enables the testing of hypotheses concerning more than two distributions. The two basic tests that will be discussed are the test for the independence of variables, and a test to determine whether a group of data assumed to be describable by a specific distribution, does in fact conform to that pattern.

Major characteristics
(a) There is no assumption of normality and it can be applied more generally.
(b) It may be utilized under conditions with a relatively small sample size.
(c) This distribution varies according to the number of degrees of freedom. For each number of degrees of freedom there is a particular distribution.

Chi-square distributions for selected degrees of freedom

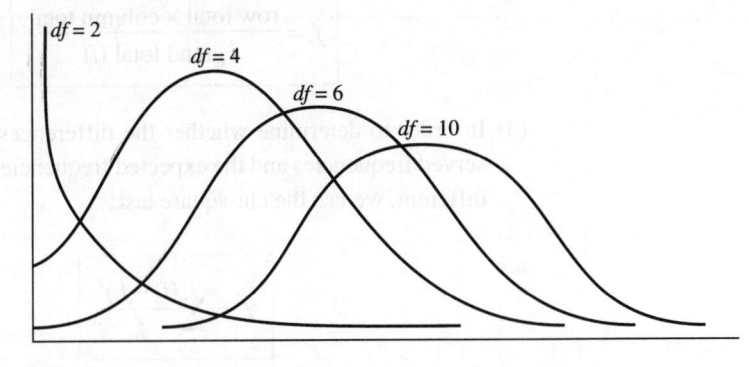

$df = 2$
$df = 4$
$df = 6$
$df = 10$

(d) The distribution is positively skewed but approaches the normal distribution as the number of degrees of freedom increases.

(e) The chi-square value is always positive.

9.2 TEST OF INDEPENDENCE

With this test, it is possible to test whether variables are independent of each other.

Procedure: (1) State H_0 and H_a:

The null hypothesis states that the two variables are statistically independent. This means that knowledge of the one variable does not help in predicting the other variable.

H_0: the variables are independent.

H_a: the variables are dependent.

(2) Since two variables are involved, the observed frequencies (f_o) are entered in a two-way classification table, or contingency table.

The dimensions of such a table are described by identifying the number of rows (r) and the number of columns (k) in the identity ($r \times k$). There may be any number of subdivisions within the two basic classifications. In order to perform the chi-square test of independence, something with which to compare the observed values is needed. The logical comparison is between observed frequencies and expected frequencies.

A table of expected frequencies (f_e) is determined based on H_0 being true. The f_e for any given cell is the product of the total of the frequencies observed (f_o) in that row and that column divided by the overall size of the sample.

$$f_e = \frac{\text{row total} \times \text{column total}}{\text{grand total } (t)}$$

(3) In order to determine whether the differences between the observed frequencies and the expected frequencies are significantly different, we use the chi-square test:

$$\chi^2 = \sum \frac{(f_o - f_e)^2}{f_e}$$

THE CHI-SQUARE TEST

Calculate the value of the chi-square by substituting cell by cell the values from the f_o and f_e tables into the formula.

(4) In order to determine whether the chi-square value is high or low (how different from the fe can the fo be and still support the H0?), it is compared with a critical chi-square value from the chi-square table. The top row of the table shows the significance level specified along with H0. The first column contains the number of degrees of freedom.

Determining the number of degrees of freedom:
= (number of rows −1) × (number of columns −1)
= $(r-1)(k-1)$

(5) *Decision:* The acceptance region for H_0 goes from the left tail of the curve to the chi-square critical value. To the right lies the rejection region. Compare the chi-square value with region of rejection and reject H_0 or do not reject H_0.

(6) State your conclusion.

Example A survey was conducted by a large mail order supplier. In a recent mailout of specials to a list of customers, a random sample of 1 200 customers received three types of response envelopes: one with prepaid postage, one without postage and a foldup type with the order form actually folded into the envelope with postage required. The purpose of this investigation is to test the relationship between response and type of envelope. ($\alpha = 0.01$)

	Prepaid postage	*No postage*	*Fold-up return*	*Total*
Response:	300	150	175	625
No response:	200	250	125	575
TOTAL:	500	400	300	1 200

H_0: response is independent of type of envelope.
H_a: response is dependent of type of envelope.
$\alpha = 0.01$

f_o		f_e	$\dfrac{(f_o - f_e)^2}{f_e}$
300	625 × 500 ÷ 1 200	= 260.42	6.016
200	575 × 500 ÷ 1 200	= 239.58	6.539
150	625 × 400 ÷ 1 200	= 208.33	16.332

f_o	f_e		$\dfrac{(f_o - f_e)^2}{f_e}$
250	575 × 400 ÷ 1 200	= 191.67	17.751
175	625 × 300 ÷ 1 200	= 156.25	2.25
125	575 × 300 ÷ 1 200	= 143.75	2.446
1 200		1 200	c^2 = 51.334

Test statistic

$$\chi^2 = \sum \dfrac{(f_o - f_e)^2}{f_e}$$

$$= 51.334$$

$$df = (r-1)(k-1) = (2-1)(3-1) = 2$$

Decision rule

Reject H_0 if χ^2-test > 9.21034

Decision: Because the test statistic 51.334 is in the rejection region, the null hypothesis should be rejected at $\alpha = 0.01$.

Conclusion: There is evidence to believe that the response rate and type of envelope are not independent.

9.3 GOODNESS-OF-FIT TESTS

These tests are used to determine whether a set of sample data follows a particular probability distribution. These tests will determine whether the sample information is significantly different from that expected for the hypothesized distribution. This could be very important in supporting the assumption necessary for other statistical techniques.

Uniform distribution This test is concerned with testing a null hypothesis that the population distribution for a random variable follows a specified form.

Example A manufacturer of toothpaste hopes to market a toothpaste with the following flavours: spearmint, lime, vanilla, strawberry and mint. He wants to test if consumers have a preference with respect to flavours or whether they like all tastes equally. Small tubes of each flavour were given to 200 customers and each was asked to state his preference. The following results were obtained:

Flavour	f_o	f_e	$\dfrac{(f_o - f_e)^2}{f_e}$
spearmint	32	40	1.6
lime	30	40	2.5
vanilla	28	40	3.6
strawberry	58	40	8.1
mint	52	40	3.6
	200	200	$\chi^2 = 19.4$

H_0: there is no preference, i.e. the data are uniformly distributed.
H_a: there is a preference in respect of flavour.
$\alpha = 0.01$
$df = k - 1 - m$
$\quad = 5 - 1 - 0$
$\quad = 4$

Where:
$\quad k$ = number of groups or intervals
$\quad m$ = number of population parameters to be estimated. (In the case of uniform testing no parameters are estimated.)

Decision rule

Reject H_0 if $\chi^2 > 13.277$

Test statistic

$$\chi^2 = \sum \dfrac{(f_o - f_e)^2}{f_e}$$
$$= 19.4$$

Decision: Because the test statistic 19.4 is in the rejection region, the null hypothesis should be rejected at $\alpha = 0.01$

Conclusion: There is evidence to suggest that there is a difference with respect to taste preference.

Binomial goodness-of-fit test A health inspector checks take-away outlets to see if they measure up to five conditions. In the most recent 200 inspections, the number of instances where conditions were not complied with, were as follows:

No of conditions (x): 0 1 2 3 4 5 Total
Instances of non-compliance (f): 20 58 64 28 20 10 200

Using $\alpha = 0.01$, test the hypothesis that the number of instances of non-compliance is a binomially distributed random variable.

- H_0: Distribution is binomial.
- H_a: Distribution is not binomial.
- $\alpha = 0.01$

  $df = k - 1 - m$
  $= 5 - 1 - 1$
  $= 3$

  ($k = 5$ and not 6 in accordance with a rule to combine adjacent classes when $f_e < 5$)

- *Decision rule*

  Reject H_0 if $\chi^2 > 11.345$.

- χ^2 test:

  The key parameter is π with the mean $= n\pi$

  $$\bar{x} = \frac{\sum fx}{n}$$

  $$= \frac{0(20) + 1(58) + 2(64) + 3(28) + 4(20) + 5(10)}{200}$$

  $$= 2$$

  mean $= n\pi$

  $$\pi = \frac{\bar{x}}{n}$$

  $$= \frac{2}{5}$$

  $$= 0.4$$

- By making use of the binomial formula and $x = 0$ to 5; $n = 5$ and $\pi = 0.4$, compute the binomial probabilities.
- There is a rule that no fe should be less than 5. When this happens, as below, combine adjacent classes.

x	fo	$P(x)$	fe	$\dfrac{(f_o - f_e)^2}{f_e}$
0	20	0.0778	15.6	1.24
1	58	0.2592	51.8	0.74
2	64	0.3456	69.1	0.38
3	28	0.2304	46.1	7.11
4	20 ⎫ 30	0.0768	15.4 ⎫ 17.4	9.12
5	10 ⎭	0.0102	2.0 ⎭	
	200	1.0000	200.0	$\chi^2 = 18.59$

THE CHI-SQUARE TEST

❏ *Decision:* Because the test statistic 18.59 is in the rejection region, the null hypothesis should be rejected at $\alpha = 0.01$.
❏ *Conclusion:* There is evidence to suggest that the distribution does not fit the binomial distribution.

Poisson goodness-of-fit test The following is the distribution of the numbers of calls received at the switchboard of an agency during 600 five-minute intervals:

Number of calls:	0	1	2	3	4	5	6	7
Frequency:	34	131	160	136	72	37	22	8

Test at a 0.01 level of significance, the hypothesis that the underlying distribution from which the sample came is a Poisson distribution with the parameter $\mu = 2.5$. (If the parameter is not given, you have to estimate it and consequently lose an additional degree of freedom.)

❏ H_0: The population sampled has a Poisson distribution.
❏ H_a: The population sampled has some other distribution.
❏ $\alpha = 0.01$

❏ *Decision rule*
  Reject H_0 if $\chi^2 > 18.475$ ($df = k - 1 - m = 8 - 1 - 0 = 7$)
❏ χ^2 test

$P(x)$	$600 \cdot P(x) = fe$	fo	$\dfrac{(fo - fe)^2}{fe}$
$P(x=0) = \dfrac{2.5^0 \cdot e^{-2.5}}{0!} = 0.082$	49.2	34	4.696
$P(x=1) = \dfrac{2.5^1 \cdot 0.082}{1!} = 0.205$	123.0	131	0.520
$P(x=2) = \dfrac{2.5^2 \cdot 0.082}{2!} = 0.256$	153.6	160	0.267
$P(x=3) = \dfrac{2.5^3 \cdot 0.082}{3!} = 0.214$	128.4	136	0.45
$P(x=4) = \dfrac{2.5^4 \cdot 0.082}{4!} = 0.133$	79.8	72	0.762
$P(x=5) = \dfrac{2.5^5 \cdot 0.082}{5!} = 0.067$	40.2	37	0.255
$P(x=6) = \dfrac{2.5^6 \cdot 0.082}{6!} = 0.028$	16.8	22	1.61
$P(x \geq 7) = 1 - P(x=0 \ldots 6)$ $= 1 - (0.985)$ $= 0.015$	9	8	0.111

$$\chi^2 = 8.671$$

(Assume the last interval to be open.)

- *Decision:* Because the test statistic 8.671 is not in the region of rejection, the null hypothesis should not be rejected at $\alpha = 0.01$
- *Conclusion:* There is evidence to suggest that the Poisson distribution provides a good fit.

Normal distribution goodness-of-fit test

Per capita income in a sample of 103 countries

Income (R)	Frequency
2 500–3 099	12
3 100–3 699	15
3 700–4 299	31
4 300–4 899	20
4 900–5 499	15
5 500–6 099	10
	103

The data in the above table are to be tested to determine if they are normally distributed.

- H_0: per capita incomes are normally distributed.
- H_a: per capita incomes are not normally distributed.
- $\alpha = 0.05$

$df = k - 1 - m = 6 - 1 - 2 = 3$

(Parameters to be estimated are μ and σ.)

- *Decision rule*

Reject H_0 if $\chi^2 > 7.815$

- *Test statistic*

χ^2 test = 3.361

(f_e is calculated by determining the probability of an observation being in an interval and multiply it by $\sum f_o$. The first and last intervals always have to be open because the total probability = 1. Remember to group some of the intervals to ensure that the f_e for each class is at least 5.)

THE CHI-SQUARE TEST

	f_o	interval $P(x)$	f_e $P.\Sigma f$	$\dfrac{(f_o - f_e)^2}{f_e}$
less than 3 099.5	12	.0968	9.97	0.413
3 099.5–3 699.5	15	.1708	17.59	0.381
3 699.5–4 299.5	31	.2603	26.81	0.655
4 299.5–4 899.5	20	.2485	25.60	1.225
4 899.5–5 499.5	15	.1487	15.32	0.0067
5 499.5 and over	10	.0749	7.71	0.680
	103	1.0000	103.0	3.361

Estimate μ and σ using $\bar{x} = 4238.3$ and $s = 874.5$ which are calculated from the frequency table.

❏ Upper boundary of first interval:

$$z = \dfrac{x - \mu}{\sigma}$$

$$= \dfrac{3\,099.5 - 4\,238.3}{874.5}$$

$$= -1.3$$

(Area = 0.4032 in normal table)

P (x < 3 099.5) = 0.5 − 0.4032

= 0.0968

❏ Upper boundary of second interval:

$$= \dfrac{3\,699.5 - 4238.3}{874.5}$$

$$= -0.62$$

(Area = 0.2324 in normal table.)

P (3 099.5 < x < 3 699.5) = 0.4032 − 0.2324

= 0.1708

❏ Third interval:

$$z = \dfrac{4\,299.5 - 4\,238.3}{874.5}$$

= 0.07

Area: 0.0279

P (3699.5 < x < 4299.5) = 0.2324 + 0.0279

= 0.02603

- Fourth interval:

$$z = \frac{4\,899.5 - 4\,238.3}{874.5}$$

$$= 0.76$$

Area: 0.2764

P (4299.5 < x < 4899.5) = 0.2764 − 0.0279

$$= 0.2485$$

- Fifth interval:

$$z = \frac{5\,499.5 - 4\,238.3}{874.5}$$

$$= 1.44$$

Area: 0.4251

P (4 899.5 < x < 5 499.5) = 0.4251 − 0.2764

$$= 0.1487$$

- Sixth interval:

P (x ≥ 5 499.5) = 0.5 − 0.4251

$$= 0.0749$$

- *Decision:* Because the test statistic 3.361 is not in the region of rejection, the null hypothesis should not be rejected at $\alpha = 0.05$
- *Conclusion:* There is evidence to suggest that the distribution is normal.

9.4 TESTS CONCERNING SMALL PROPORTIONS

As explained in a previous chapter, the normal distribution can be used to test hypotheses concerning proportions when np and $nq \geq 5$. If the sample size is small and a normal distribution can't be assumed, the χ^2 distribution can be used.

Test concerning one proportion

Suppose it was hypothesized that a proportion of $\pi = 0.30$ of men will react positively to a new disposable razor. A random sample of 20 men is selected and two of the men have a positive reaction to the new razor and 18 don't.

THE CHI-SQUARE TEST

	f_o	f_e	
Positive	2	6	(0.3×20)
Negative	18	14	(0.7×20)
	20	20	

$H_0: \pi = 0.30$

$H_a: \pi \neq 0.30$

$\alpha = 0.05$

$df = k - m - 1 = 2 - 0 - 1 = 1$

Reject H_0 if $\chi^2 > 3.843$

Test statistic

$$\chi^2 = \sum \frac{(|f_o - f_e| - 0.5)^2}{f_e}$$

$$= \frac{(|2-6| - 0.5)^2}{6} + \frac{(|18-14| - 0.5)^2}{14}$$

$$= 2.917$$

Note: The 0.5 adjustment in the formula is an adjustment required when discrete data and a continuous distribution are used. It is known as 'correction for continuity'. The adjustment is always one half of the nearest unit of measure.

Decision: H_0 cannot be rejected.

Conclusion: The proportion of men that will react positively to a new disposable razor is 0.30.

EXERCISES

Exercise 1 A firm markets its product using door-to-door methods. Recently, the sales manager has become interested in the number of sales calls made by each of the employees. He reasoned that if all the employees are working hard, they should make the same number of calls during a set period of time. In order to investigate this hypothesis, the sales manager uses a sample of 5 employees.

Employee:	A	B	C	D	E
Number of calls:	31	62	59	40	58

At the 1 % level of significance, is the manager's idea supported?

Exercise 2 A city has three television stations, each of which has its own evening news programme. The Mullin Survey group wants to determine if there is a preference among the viewing audience for any station. A random sample of 150 viewers revealed that 53 watched the evening news on A, 64 on B and 33 on C. At a 5 % level of significance, is there sufficient evidence to show that the three stations do not have equal shares of the evening news audience?

Exercise 3 Suppose that the following were the respective car manufacturers' shares of the national market.

Manufacturer	Number of cars	% of total
General Motors	22 500	45
Ford	17 500	35
Chrysler	7 500	15
Toyota	2 500	5
TOTAL:	50 000	100

A sample of 2 000 owners in a big city revealed this ownership pattern: General Motors 858; Ford 687; Chrysler 300; Toyota 155. Does the ownership pattern in the city differ from the national pattern? Use $\alpha = 0.01$.

Exercise 4 National figures revealed that 42 % of the vacationers who travel outside South Africa go to Europe, 20 % to the Far East, 16 % to America, 6 % to the Middle–East, 12 % to the Indian Ocean Islands, and 4 % elsewhere. A local travel agency wondered if its customers differ significantly from this breakdown with respect to their travel destination. A sample of 200 customers revealed:

Destination	Number
Europe	80
Far East	44
America	34
Middle East	16
Indian Ocean Islands	20
Other	6

Set $\alpha = 0.05$.

Exercise 5 A random sample of 200 residents were asked their opinion on capital punishment and their marital status. The results were as follows:

THE CHI-SQUARE TEST

Opinion	Married	Not Married
Favour	100	20
Oppose	50	30

Is there a relationship between the two variables at a 1 % level of significance?

Exercise 6 A group of executives were classified according to total income and their age. Test the hypothesis, at the 1 % level, that age is not related to level of income.

Age	Up to R100 000	Up to R400 000	Over R400 000
Under 40	6	9	5
Under 54	18	19	8
55 and older	11	12	17

Exercise 7 Louis Armstrong, salesman for Dillard Paper Company, has 5 clients to visit per day. It is suggested that this variable is described by the binomial distribution, and that the probability of selling to each client is 0.3. Given the following observed frequencies of Mr Armstrong's sales per day, can we conclude that the distribution does in fact follow the suggested distribution? Use the 0.05 level of significance.

No of sales per day	0	1	2	3	4	5
f_o	20	65	42	14	6	3

Exercise 8 Below is an observed frequency distribution with $\mu = 2.44$ and $\sigma = 0.4$. At the 0.01 level of significance, does this distribution seems to be well described by a suggested normal distribution?

Interval	f
less than 1.8	3
1.8–2.19	17
2.2–2.59	33
2.6–2.99	22
3.0 or more	5

Exercise 9 A large city fire brigade calculates that for any given area, during any given 8-hour shift, there is a 30 % chance of receiving at least one fire alarm. Here is a random sample of 60 days:

No. of shifts:	0	1	2	3
No. of alarms:	16	27	11	6

At the 0.05 level of significance, do these fire alarms follow a binomial distribution?

Exercise 10 A hospital administrator has examined past records from 300 randomly selected 8-hour shifts to determine the frequency with which the hospital treats fractures. The number of days in which 0, 1, 2, 3, 4, 5, 6 or more patients with broken bones were treated was 25, 45, 63, 71, 48, 26 and 22 respectively. At the 5 % level of significance, can we reasonably believe that the incidence of broken bone cases follows a Poisson distribution with $\mu = 3$?

CHAPTER 10
Analysis of Variance (ANOVA)

CONTENTS

10.1 INTRODUCTION 173
10.2 PROCEDURE FOR ANOVA 174
10.3 MULTIPLE PAIRWISE COMPARISON OF MEANS . . 177
10.4 HOMOGENEITY OF VARIANCES 180

10.1 INTRODUCTION

The analysis of variance technique tests the hypothesis that more than two samples have the same means. It is also possible to use ANOVA for a small sample situation, since the F-distribution changes according to the number of degrees of freedom.

Assumptions made in ANOVA

(1) Each of the samples is drawn from a normal population. If the samples are large, the central limit theorem can be used as the basis for normality.
(2) The samples are independent and selected at random.
(3) Each population has the same variance (σ^2). If there are equal sample sizes in each group, inferences based on the F-distribution may not be seriously affected by unequal variances. However, if the sample sizes are not the same for each group, unequal variances from group to group can have a serious effect on drawing inferences made from the ANOVA. Later in the chapter a discussion will follow on procedures that have been developed to test the assumption of homogeneity of variances.

10.2 PROCEDURE FOR ANOVA

(1) State the null (H_0) and alternative (H_a) hypothesis.

H_0: $\mu_1 = \mu_2 = \mu_3 \ldots$

H_a: One or more of the population means is not equal to the others.

(2) Specify the level of significance — a criterion for rejection of the H_0 is necessary.

(3) Use the F-distribution to identify the acceptance and rejection region for H_0. The F-distribution is a whole family of positive skewed distributions and each is identified by a pair of degrees of freedom. The first number refers to the number of degrees of freedom in the numerator of the F-ratio and the second to the degrees of freedom in the denominator. This test, or ratio, is based on a comparison of two different ways to estimate the overall population variance:

- The between group variance (V_B) measures the differences between the sample mean of each group (x) and the grand or overall mean $\bar{\bar{x}}$
- The within group variance (V_W) measures the variation of each value from the mean of its own group and averages these differences over all groups.

df in the numerator = Number of samples $- 1$

$= k - 1$

df in the denominator = Total sample size $- k$

$= \sum n - k$

The distribution tends to get more and more normal as the number of degrees of freedom increase.

The critical F-value is obtained from the F-table. Separate tables exist for each level of significance. The columns represent the number of degrees of freedom for the numerator and the rows represent the number of degrees of freedom for the denominator. The table value sets the upper limit of the acceptance region.

(4) Compute the value of the test statistic by making use of the ANOVA technique.

$$F\text{-ratio} = \frac{V_B}{V_W}$$

With:

V_B = estimate of population variance based on variance between sample means

ANALYSIS OF VARIANCE (ANOVA)

V_W = estimate of population variance based on variance within samples

(i) Set up the table of observations.
(ii) Complete a mean ($\bar{x}$) for each sample.
(iii) Calculate a 'total' mean ($\bar{\bar{x}}$)

$$\bar{\bar{x}} = \frac{\text{Total of all sample items}}{\text{Number of sample items}}$$

(iv) Calculate the standard deviation (s) for each sample and square the answer. (Variance = s^2)

(v) Determine

$$V_B = \frac{\sum n(\bar{x} - \bar{\bar{x}})^2}{k - 1}$$

(vi) Determine

$$V_W = \frac{\sum (n - 1) s^2}{\sum n - k}$$

(5) Decision

If the F-ratio is not larger than the table F-value, then the conclusion is that there is not sufficient evidence to indicate that one or more of the population means are not equal to the others. If the calculated F-ratio is larger than the table F-value, then the conclusion is that one or more of the population means is not equal to the others.

The denominator (V_W) and numerator (V_B) should be about equal if the H_0 is true. The nearer the F-ratio comes to one, the greater the inclination to accept H_0. Conversely, as the F-ratio becomes larger, the inclination to reject H_0 will increase — it will become more and more likely that a difference does exist between one or more of the population means. The F-ratio will vary from 1 by chance alone if the samples came from the same normally distributed population with a mean and variance of μ and σ^2. If, however, the sample data came from two or more normal populations that have different means and a common variance, the F-ratio will tend to be greater than one. If the F-ratio is smaller than one, it is most probably because the assumption of equal variances is not true.

Example A grocery chain consisted of 6 stores. Three years ago the chain acquired two other groups of stores. One of these contained 4 stores and the other contained 5 stores. Turnover of salespeople is a problem at all 15 stores. The number of salespeople who resigned at each of the stores last year is shown in the table. The employee compensation and benefit policies of the three groups of stores have not been standardized. The personnel manager of the entire 15 store chain is interested in knowing whether there is a real difference in the turnover of the personnel in the three groups of stores. If there is, he will consider standardizing compensation and benefits.

	Group 1	Group 2	Group 3
	10	6	14
	8	9	13
	5	8	10
	12	13	17
	14		16
	11		
n:	6	4	5
$\bar{x}$:	10	9	14
s^2:	10	8.67	7.5

Grand mean ($\bar{\bar{x}}$) = 11.07

H_0: $\mu_1 = \mu_2 = \mu_3$
H_a: The mean annual turnover is not the same for all three groups of stores.
$\alpha = 0.05$

Decision rule:

Reject H_0 if F-ratio > 3.89
$V_1 = k - 1 = 3 - 1 = 2$
$V_2 = \Sigma n - k = 15 - 3 = 12$

$$V_B = \frac{\Sigma n(\bar{x} - \bar{\bar{x}})^2}{k - 1}$$

$$= \frac{6(10 - 11.07)^2 + 4(9 - 11.07)^2 + 5(14 - 11.07)^2}{3 - 1}$$

$$= 33.47$$

$$V_W = \frac{\Sigma(n - 1)s^2}{\Sigma n - k}$$

ANALYSIS OF VARIANCE (ANOVA)

$$= \frac{(6-1)10 + (4-1)\,8.67 + (5-1)\,7.5}{6+4+5-3}$$

$$= 8.83$$

$$\text{F-ratio} = \frac{V_B}{V_W}$$

$$= \frac{33.47}{8.83}$$

$$= 3.79$$

Decision: Do not reject H_0.

Conclusion: On the basis of these data, the personnel manager cannot expect to reduce turnover by standardizing compensation and benefits.

10.3 MULTIPLE PAIRWISE COMPARISON OF MEANS

If the H_0 of equal means has been rejected, we may desire to know which means are not equal in order to decide which group has the higher or lower mean value. Such an examination is made by multiple pairwise comparison of means. Several comparison tests are available. Two of these will be discussed:

Sheffe's test This is one of the most flexible and conservative tests available. It requires only the use of the F-tables. The absolute difference between two sample means must exceed the critical difference (CD) to be declared significant.

Steps:

(1) Prepare a matrix showing the mean of each sample in ascending order of magnitude and the absolute difference between pairs of means.

(2) Find the F_α value from the F-table using $df_1 = k - 1$ and $df_2 = \sum n - k$

(3) Find CD

$$CD = \sqrt{(k-1)\,(F_{\alpha;k-1;\sum n - k})} \cdot \sqrt{V_W \cdot \left[\frac{1}{n_i} + \frac{1}{n_j}\right]}$$

where n_i and n_j are the sizes of the two samples

(4) Compare each absolute difference with the CD. If the absolute difference exceeds CD, we can conclude that the difference between that pair of means is significant.

Example   Refer to the example of the grocery chain store, with:
$$\bar{x}_1 = 10; \quad \bar{x}_2 = 9; \quad \bar{x}_3 = 14$$
$$n_1 = 6; \quad n_2 = 4; \quad n_3 = 5$$
$$V_W = 8,83$$

- Matrix:

	$\bar{x}_2$	$\bar{x}_1$	$\bar{x}_3$
$\bar{x}_2 = 9$	–	1	5
$\bar{x}_1 = 10$	–	–	4
$\bar{x}_3 = 14$	–	–	–

- $F_{\alpha;k-1;\sum n-k} = F_{0,05;\ 3-1;\ 15-3}$
$$= 3.89$$

For H_0: $\mu_1 = \mu_3$
$$CD = \sqrt{2 \times 3.89} \cdot \sqrt{8.83(1/6 + 1/5)}$$
$$= 5.02$$

Since $4 < 5.02$ do not reject H_0: The difference between μ_1 and μ_3 is not significant.

For H_0: $\mu_2 = \mu_3$
$$CD = \sqrt{2 \times 3.89} \cdot \sqrt{8.83(1/4 + 1/5)}$$
$$= 5.56$$

Since $5 < 5.56$ do not reject H_0: The difference between μ_2 and μ_3 is not significant.

For H_0: $\mu_1 = \mu_2$
$$CD = \sqrt{2 \times 3.89} \cdot \sqrt{8.83(1/6 + 1/4)}$$
$$= 5.35$$

Since $1 < 5.35$ do not reject H_0: The difference between μ_1 and μ_2 is not significant.

Tukey's method   Tukey (1953) has developed such a test which he named the HSD (honestly significance difference) test. A difference between two means is significant at a given a-level if it equals or exceeds HSD which is:

$$\boxed{HSD = q_{a,k,\sum n-k} \cdot \sqrt{\frac{V_W}{n_{smallest}}}}$$

ANALYSIS OF VARIANCE (ANOVA)

Where: V_W = estimate of the within group variance
$n_{smallest}$ = the smallest of the sample sizes obtained
$q_{\alpha,k,\sum n-k}$ = obtain from table: percentage points of the studentized range

(1) Prepare a matrix showing the mean of each sample in ascending order of magnitude and the absolute difference (d) between pairs of means.
(2) Determine the q_α value with $\sum n - k$ degrees of freedom and k number of means.
(3) Determine HSD.
(4) Compare each absolute difference with the HSD. If the difference exceeds HSD, it can be concluded that those differences are statistically significant.

Example Refer to example of the grocery chain store:
Matrix:

$$\begin{array}{cccc} & \bar{x}_2 & \bar{x}_1 & \bar{x}_3 \\ \bar{x}_2 = 9 & - & 1 & 5 \\ \bar{x}_1 = 10 & - & - & 4 \\ \bar{x}_3 = 14 & - & - & - \end{array}$$

$q_{\alpha,k,\sum n-k} = q_{0.05,3,12} = 3.77$

$$HSD = q_x \cdot \sqrt{\frac{V_W}{n_{smallest}}}$$

For $H_0: \mu_1 = \mu_3$

$$HSD = 3.77 \cdot \sqrt{\frac{8.83}{5}}$$

$= 5.01$

Since 4 < 5.01 do not reject H_0, the difference between the means is due to chance.

For $H_0: \mu_2 = \mu_1$

$$HSD = 3.77 \cdot \sqrt{\frac{8.83}{4}}$$

$= 5.6$

Since 1 < 5.6 do not reject H_0, the difference between the means is due to chance.

For $H_0: \mu_2 = \mu_3$

$$HSD = 3.77 \cdot \sqrt{\frac{8.83}{4}}$$

$$= 5.6$$

Since 5 < 5.60 do not reject H_0, the difference between the means is due to chance.

10.4 HOMOGENEITY OF VARIANCES

Where there is concern that the assumption of equal variances may be unrealistic, the validity can be checked.

Hypothesis testing for the equality of variances from two populations

In order to examine the equality of two variances of two independent populations, a statistical procedure has been developed that is based upon the ratio of the two sample variances. If the data from each population are assumed to be normally distributed and all possible samples of size n_1 and n_2 are selected, we can calculate s_1^2 and s_2^2 for each of these samples. Suppose we form all possible ratios of $\frac{s_1^2}{s_2^2}$, the result will be known as the F-distribution.

(1) $H_0: \sigma_1^2 = \sigma_2^2$
(2) $H_a: \sigma_1^2 \neq \sigma_2^2$
(3) Select level of significance.
(4) Decision rule: Form a rejection area for H_0 using a critical F-value located in the F-table. Each F-value has two different degrees of freedom associated with it.

$df_1 = n_1 - 1$ number of degrees of freedom in the larger sample variance

$df_2 = n_2 - 1$ number of degrees of freedom in the smaller sample variance

When forming the rejection area we want to make certain that the upper tail is used because only the upper-tail values of F are shown in the table. To accomplish this, the larger sample variance will always be placed in the numerator of the F-ratio.

ANALYSIS OF VARIANCE (ANOVA)

Either one-tailed or two-tailed tests can be employed:

When a two-tailed test is to be done, the critical value will be $F\left(df_1, df_2, \dfrac{\alpha}{2}\right)$. Only the right-tail critical value will be needed.

When a one-tailed test, with the critical region on the left, is to be done, the position of the two variances will be interchanged in the statement of hypothesis — this will reverse the direction of H_a and put the critical region on the right with $F(df_1, df_2, \alpha)$.

(5) Test statistic:

$$\text{F-ratio} = \frac{\text{larger sample variance}}{\text{smaller sample variance}}$$

$$= \frac{s_1^2}{s_2^2} \text{ or } \frac{s_2^2}{s_1^2}$$

(6) Decision: Compare F-ratio, which is the test statistic, with the F-value from the tables.

If F-ratio > F-value, reject H_0 and conclude that the population variances differ significantly. If F-ratio < F-value, conclude that insufficient sample evidence exists to reject H_0, and that the difference that does exist is a chance difference.

Example A manufacturer believes he has two types of tyres (assume tyre wear to be normally distributed) that have the same variability in wear. If a sample of 25 tyres of the first type produced a variance of 2 000 km² and a sample of 20 tyres of the second type produced a variance of 1 600 km², is the manufacturer correct?

$H_0: \sigma_1^2 = \sigma_2^2$

$H_a: \sigma_1^2 \neq \sigma_2^2$

$\alpha = 0.10$

Rejection area: Reject H_0 if F-ratio > 2.11; $df_1 = 25 - 1$; $df_2 = 20 - 1$

$$\text{F-ratio} = \frac{s^2 \text{ larger}}{s^2 \text{ smaller}}$$

$$= \frac{2\,000}{1\,600}$$

$$= 1.25$$

Decision: Do not reject H_0
Conclusion: The difference in the variances is a chance difference.

EXERCISES

Exercise 1 The matron at a hospital has the responsibility for assigning personnel to the emergency room. Present policy calls for the same number of nurses to be assigned to all three shifts. The matron, however, thinks that the number of emergencies handled may not be the same for each shift. Use the ANOVA technique to investigate whether the same number of emergencies are handled on each shift. Which shift has the highest number of emergencies? ($\alpha = 0.05$)

A random sample of 5 days from each shift is selected and the results are shown below.

Day	Afternoon	Night
44	33	39
53	42	24
56	15	30
49	30	27
38	45	30

Exercise 2 Energy shortages have caused many schools to turn down the heat in the classrooms. The principal of Penn School is concerned that this may have an effect on achievement. To investigate this question, students from one class were randomly assigned to three groups. The groups were then placed in rooms having different temperatures. Each group received televised instruction in long division. At the end of the lesson, the same ten-question examination was given to all three groups. The results, scoring the number of correct answers out of ten, were:

16 °C: 3 5 4 3 4
18 °C: 7 6 8 9 6 8 8
22 °C: 4 6 5 7 6 5 4 3

(a) Is there a significant difference in the mean scores at the 0.05 level of significance?

(b) If there is a difference between the means, determine where the difference lies.

(c) Is there evidence of a difference in the variances of the three groups?

Exercise 3 Deal-on-Wheels, a manufacturer of mobile homes, is interested in the ages of buyers of each of five available floor designs. He suspects that certain designs tend to appeal to younger buyers more than others do. A random sample of 29 records is selected from last year's sales files, and the respective ages of the principal buyers are recorded. The outcome is:

A:	30	32	31	41	35	34	
B:	48	52	45	38	34	42	
C:	54	60	56	50	47	57	63	57
D:	52	50	43	42			
E:	44	48	50	41	35		

(a) Is there a difference in age of people interested in different designs? ($\alpha = 0.05$)

(b) If there is a difference, where does the difference lies?

Exercise 4 A social worker wants to evaluate the ethical behaviour of attorneys in four different towns. Samples are selected and an index is compiled of the ethical behaviour of each attorney in the study. At a 1 % level of significance, is there a statistical difference in ethical behaviour between the four towns?

A:	12	16	12	14	26	
B:	19	20	18	9	22	19
C:	34	29	31	19	26	
D:	19	21	17	24		

Exercise 5 The manager of Think Food grocery stores is analysing the effect of various placements of candy displays within the company's stores. He decides to conduct an experiment by locating the display in different areas in each of four mart outlets. The amount in kilograms sold in the various locations each week is recorded. Is there sufficient evidence to indicate that there is a difference in sales at the various locations? Use $\alpha = 0.05$.

Front of store near bread	Top shelf near cookies	End of aisle near meat	Third shelf near cooldrinks
76	73	89	96
75	70	82	92
83	81	85	104
87	78	79	89
81	76	80	94

Exercise 6 The following data represent yields in bags per hectare for two wheat varieties. Each was tried on 8 different farms. Test the hypothesis that

the two populations have the same variance.
Let $\alpha = 0.02$

A:	81.2	72.6	56.8	76.9	42.5	49.6	62.8	48.2
B:	56.6	58.6	45.4	39.1	42.8	65.2	40.7	49.9

Exercise 7   Two different package designs were tested for a product. The shelf life was measured for a sample of 25 packages of the first design, and a variance of 720 h^2 was recorded. The shelf life of a sample of 25 packages of the second design yielded a variance of 1 455 h^2. Test the hypothesis that the two package designs have the same variance of shelf life. Use $\alpha = 0.05$.

Exercise 8   Three packaging materials were tested for moisture retention by storing the same food product in each of them for a fixed period of time and then determining the moisture loss. Each material was used to wrap 10 food samples. The results are given in the accompanying table. Can the hypothesis that the materials are equally effective be rejected? If it is rejected, where does the difference lie? Let $\alpha = 0.05$.

	Material I	Material II	Material III
No. of packages	10	10	10
Mean loss	224	232	228
Sample variance	30	40	38

Exercise 9   The following are the number of words per minute which a secretary typed on several occasions on four different typewriters:

Typewriter C:	71	75	69	77	61	72	71	78
Typewriter D:	68	71	74	66	69	67	70	62
Typewriter E:	75	70	81	73	78	72		
Typewriter F:	62	59	71	68	63	65	72	60

Perform an analysis of variance to test at the 0.01 level of significance whether the differences among the four sample means can be attributed to chance. If not, where do the differences occur?

Exercise 10   In order to test the reliability of four brands of tyres, you buy 3 sets and use them over a period of time. The life (thousands of kilometres) of each of the three sets of four brands is shown in the table.

Set	Brand A	Brand B	Brand C
A	25	12	20
B	20	9	17
C	15	6	22

Would you assume that the average life of the four brands is the same? Use the 0.01 significance level.

CHAPTER 11
Simple Regression and Correlation

CONTENTS

11.1 INTRODUCTION 187
11.2 LINEAR REGRESSION ANALYSIS 187
11.3 CORRELATION ANALYSIS 188
11.4 STEP-BY-STEP PROCEDURE 188
11.5 SPEARMAN'S RANK-ORDER
　　　CORRELATION COEFFICIENT 195

11.1 INTRODUCTION

Regression and correlation analysis are statistical tools used to study the relationship between two or more variables of which one is dependent and the other independent. It is used to determine whether there is a relationship between the variables: how good that relationship is; and how the relationship can be used to make estimates with some accuracy. For our purposes, it will be assumed that the relationship between the variables approaches a straight line graphically.

It is extremely important to understand the meaning of these coefficients and be able to test their statistical significance. Otherwise, the computation of estimates is of little value.

11.2 LINEAR REGRESSION ANALYSIS

The general process of predicting one dependent variable (y) based on an independent variable (x), is known as regression. The main objective is to develop an estimating equation or mathematical formula that relates the known variable to the unknown estimated variable.

$$\hat{y} = a + bx$$

where x and y are variables and a and b are parameters to be estimated.

11.3 CORRELATION ANALYSIS

Correlation analysis is used to describe the degree of strength by which one variable is linearly related to another. This measure is based on a scale between -1 and +1. If there is no relationship between x and y, then $r = 0$. The strength of the correlation is not dependent on direction. Coefficients of 0.95 and –0.95 are equal in strength.

11.4 STEP-BY-STEP PROCEDURE

1   Collect data for both dependent (y) and independent (x) variables

2   Arrange the data in two columns, x and y.

3   Draw a scatter diagram of the data points that represent graphically the relationship between x and y. This will illustrate whether there is a relationship and also suggest whether it is linear, positive or negative. The strength of the relationship may be concluded tentatively. A scatter diagram consists of coordinate points of the two variables plotted on a graph. The points are not connected. The value of the independent x is plotted in respect of the horizontal axis, and the value of the dependent y is plotted in respect of the vertical axis.

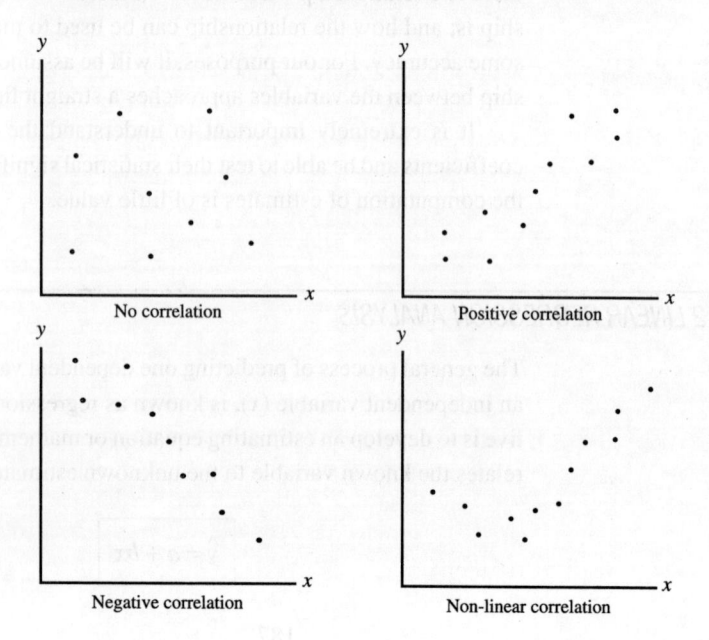

No correlation

Positive correlation

Negative correlation

Non-linear correlation

SIMPLE REGRESSION AND CORRELATION

4. Compute Pearson's coefficient of correlation (r): The coefficient of correlation measures the degree of association between the variables. When the slope of the estimating equation is negative, r is negative; and when it is positive, r is positive. Thus, the sign of r indicates the direction of the relationship between the two variables x and y. If an inverse relationship exists, that is, if y decreases as x increases, then r will fall between 0 and -1. If there is a direct relationship, that is, if y increases as x increases, then r will fall between 0 and $+1$. This measure enables us to make statements such as: the correlation is strong, weak, etc.

Steps
- Compute the additional columns, xy, x^2 and y^2, and sum all the columns.
- Substitute the appropriate values into the formulae for r.

$$r = \frac{n \cdot \sum xy - \sum x \cdot \sum y}{\sqrt{\left[n \cdot \sum x^2 - (\sum x)^2\right]\left[n \cdot \sum y^2 - (\sum y)^2\right]}}$$

- Interpret the answer.

5. Compute the coefficient of determination (r^2):
This coefficient does have a more precise meaning — it is the primary way of measuring the strength of the association that exists between two variables. It is developed from the relationship between the variation of the given y-values around the fitted regression line and their own mean line. Variation is measured by the sum of a group of squared deviations.

$$r^2 = 1 - \frac{\text{variation of } y\text{-values around regression line}}{\text{variation of } y\text{-values around their own mean}}$$

$$= \text{variation of } y - \text{values that are explained by } \hat{y}$$

This is a more conservative measure of the relationship because it enables us to calculate the total variation in y that is explained by x. To determine the value of this measure, simply square the coefficient of correlation and multiply the answer by 100 to enable you to express it as a percentage. The answer will always be positive within the range 0–100 %.

6   Testing the significance of the coefficient of correlation
    Due to the small sizes of the samples, the question arises whether it is possible that the correlation in the population is really 0 and the apparent relationship is due to chance. To test this, the normal hypothesis procedure is applied:
    - H_0: population correlation coefficient (ρ) = 0
    - H_a: population correlation coefficient (ρ) ≠ 0
    - Select the level of significance.
    - Formulate the decision rule. The appropriate test statistic is the t-distribution connected with n – 2 degrees of freedom. A two-tailed test is normally applied.
    - Compute t-test:

    $$t = \frac{r \cdot \sqrt{n-2}}{\sqrt{1-r^2}}$$

    - Based on the decision rule, either reject or accept the null hypothesis.

7   Make predictions by making use of either the regression equation or the straight line graph:
    The basic idea of regression is finding a straight line that best fits the sets of data points. The Least Squares Principle is a method used to determine the regression equation by minimizing the sum of the squares of the differences between the actual y-value and the predicted values of y.
    Regression equation:

    $$\hat{y} = a + bx$$

    Where:
    $\hat{y}$ = the estimated y-variable for the value of the x-variable
    a = the interception on the y axis
    b = the slope (the average change in $\hat{y}$ for each change of 1 unit in x)

    $$b = \frac{n \cdot \sum xy - \sum x \sum y}{n \cdot \sum x^2 - (\sum x)^2}$$

SIMPLE REGRESSION AND CORRELATION

$$a = \bar{y} - b\bar{x}$$

- Compute the additional columns, xy, x^2 and y^2, and sum all the columns.
- Substitute the appropriate values into the formulae for a and b.
- Substitute the a and b values into the regression equation:
$$\hat{y} = a + bx$$
- For equation predictions, substitute a value in the place of the independent x to estimate a value for $\hat{y}$.
- Fitting a straight line to the data:
The best fitting straight line to the data can be obtained by making use of the regression equation which is a mathematical function for a straight line. When the actual dependent variable y is plotted against x, the result is a scatter diagram. When the estimated $\hat{y}$ is plotted against x and the points are connected, the result is called the regression line.

Interpretations:

The slope b tells us the amount of increase or decrease in $\hat{y}$ (depending on the sign) that will occur for a one unit increase in x, the independent variable.

The intercept a indicates the value of $\hat{y}$ when $x = 0$.

8 The standard error of estimate

The standard error of estimate was developed to measure the reliability of the estimating equation. It measures the variability, or scatter, of the observed values around the regression line. The more widely scattered these points, the larger the value of the standard error. If $Se = 0$, the estimating equation is a perfect estimator of the dependent variable. In that case, all data points should lie directly on the regression line and no points will be scattered around it.

$$Se = \sqrt{\frac{\sum y^2 - a\sum y - b\sum xy}{n - 2}}$$

9 **Approximate prediction intervals**

The standard error of estimate can be used to make a probability statement about the interval around an estimated value of $\hat{y}$ within which the actual value of y lies. These intervals are known as prediction intervals. They serve the same function as the confidence intervals. The formula for these intervals for an individual value of y is:

$$\hat{y} \pm t \cdot Se \cdot \sqrt{1 + \frac{1}{n} + \frac{(x - x_o)^2}{\sum x^2 - n\bar{x}^2}}$$

Where:
t = value of t from the Student-t table. The degrees of freedom are $n - 2$.
x_o = the specific value of x at which we want to predict the value of $\hat{y}$.

Example A personnel trainee was studying the relationship between the number of years of employment with the company and the number of days absent from work last year. The following information was obtained from the company's records:

Employee:	A	B	C	D	E	F
Length of employment:	1	5	2	4	4	3
Absences (days):	8	1	7	3	2	4

Solution:

x	y	xy	x^2	y^2
1	8	8	1	64
5	1	5	25	1
2	7	14	4	49
4	3	12	16	9
4	2	8	16	4
3	4	12	9	16
19	25	59	71	143

$\bar{x} = 3.17$
$\bar{y} = 4.17$

(i) Scatter diagram and straight line fitting

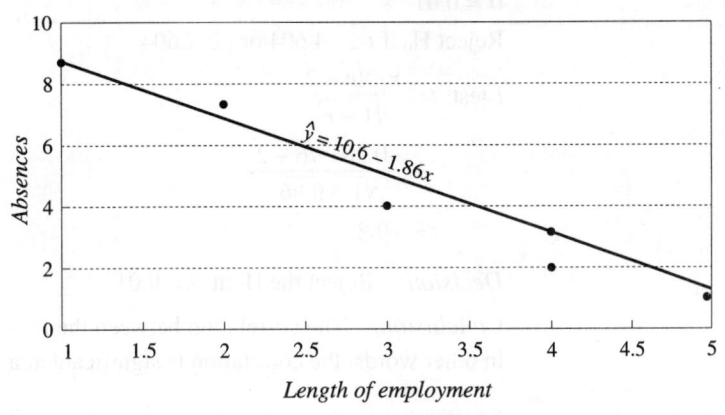

Number of years of employment and number of days absent

[Scatter diagram showing Absences (y-axis, 0 to 10) vs Length of employment (x-axis, 1 to 5), with fitted line $\hat{y} = 10.6 - 1.86x$]

From the scatter diagram we can see that there is a definite linear negative relationship between the two variables.

The linear equation

(ii) Coefficient of correlation:

$$r = \frac{n \cdot \sum xy - \sum x \cdot \sum y}{\sqrt{[n \cdot \sum x^2 - (\sum x)^2][n \cdot \sum y^2 - (\sum y)^2]}}$$

$$= \frac{6 \cdot 59 - 19 \cdot 25}{\sqrt{[6 \cdot 71 - 19^2][6 \cdot 143 - 25^2]}}$$

$$= -0.98$$

Interpretation: A very strong negative relationship exists between years of employment and days absent. In other words, the number of days absent decreases as duration of employment increases.

(iii) Coefficient of determination:
$r^2 = -0.98^2$
$= 0.96$

Interpretation: 96 % of the total variation in the number of days absent is accounted for or explained by the number of years service.

(iv) Significance testing for r:

H_0: Population correlation coefficient = 0
H_a: Population correlation coefficient $\neq 0$
$\alpha = 0.01$
Reject H_0 if $t \leq -4.604$ or $t \geq 4.604$

t-test: $t = \dfrac{r \cdot \sqrt{n-2}}{\sqrt{1-r^2}}$

$= \dfrac{-0.98 \cdot \sqrt{6-2}}{\sqrt{1-0.96}}$

$= -9.8$

Decision: Reject the H_0 at $\alpha = 0.01$

Conclusion: The correlation between the two variables is not zero, in other words, the correlation is significant at a 0.01 level.

(v) How many days would you expect a person to be absent from work if he has been in the firm's employ for four years?

Regression equation:

$b = \dfrac{n\sum xy - \sum x \sum y}{n\sum x^2 - (\sum x)^2}$

$= \dfrac{6(59) - 19(25)}{6(71) - 19^2}$

$= -1.86$

$a = \bar{y} - b\bar{x}$

$= 4.17 - (-1.86)3.17$

$= 10.06$

$\hat{y} = a + bx$

$= 10.06 - 1.86x$

Replace x with 4:

$\hat{y}_4 = 10.06 - 1.86(4)$

$= 2.62$ days

To use the line graph, the line of best fit must first be drawn using the regression equation. Determine at least two coordinate points and draw a straight line between them. Look up the x-value on the x-axis and read the estimate from the y-axis.

Regression line:
$\hat{y} = 10.06 - 1.86x$

SIMPLE REGRESSION AND CORRELATION

Replace x with 4:

$\hat{y} = 10.06 - 1.86(4)$

$= 2.62$

(4; 2.62)

Replace x with 1:

$\hat{y} = 10.06 - 1.86(1)$

$= 8.2$

(1; 8.2)

(Refer to the straight line on the graph above.)

(vi) Standard error of estimate:

$$Se = \sqrt{\frac{\sum y^2 - a\sum y - b\sum xy}{n-2}}$$

$$= \sqrt{\frac{143 - 10.06\,(25) - (-1.86)\,59}{6-2}}$$

$$= 0.56$$

The standard error is very small and it can be assumed that the estimating equation will be a good estimator.

(vii) Prediction interval:

$$\hat{y} \pm t \cdot Se \cdot \sqrt{1 + \frac{1}{n} + \frac{(x - x_0)^2}{\sum x^2 - n\bar{x}^2}}$$

$$= 2.62 \pm 2.776(0.56) \cdot \sqrt{1 + \frac{1}{6} + \frac{(3.17 - 4)^2}{71 - 6(3.17)^2}}$$

$$= 2.62 \pm 1.72$$
$$= 0.9 < \hat{y} < 4.34$$

There is a 95 % probability that a person with 4 years' service will be absent from work for between 0.9 and 4.34 days.

11.5 SPEARMAN'S RANK-ORDER CORRELATION COEFFICIENT

This coefficient measures the strength of the relationship between two variables on the basis of their ranks instead of other values. It can be interpreted in a manner similar to the simple correlation coefficient. It has the advantage that it can be used in problems where items can be ranked even though they cannot be measured on a numerical scale.

The rank-order correlation coefficient is relatively easy to compute, but should not be used in preference to the simple correlation coefficient when the original data are available. A great deal of information gets lost, and it provides a less reliable result.

Steps in calculating r'

- Rank the x and the y variables. This is accomplished by arranging the data in order and numbering them. Remember to use the same type of ranking for x and y—from high to low or from low to high. If two values are the same, they are first assigned ranks (say 2 and 3) and then the average of the ranks is determined (2.5). That average is then assigned to each appropriate value.
- Calculate the difference between ranks of the two variables for each set of values (d).
- Square these differences (d^2).
- Sum these squared differences and apply the following formula to compute the rank-order correlation coefficient.

$$r' = 1 - \frac{6\sum d^2}{n(n^2 - 1)}$$

- Interpret the coefficient.

Example

x	y	x-rank	y-rank	d	d^2
1	8	1	6	-5	25
5	1	6	1	5	25
2	7	2	5	-3	9
4	3	4.5	3	1.5	2.25
4	2	4.5	2	2.5	6.25
3	4	3	4	-1	1
					68.5

$$r' = 1 - \frac{6(68.5)}{6(6^2 - 1)}$$

$$= -0.96$$

There is very good negative correlation between the two variables.

EXERCISES

Exercise 1 The owner of several farms in different parts of the country has been studying the relationship between the July rainfall (in inches) and the yields of maize (in bags) for plots of land of the same size. The following sample information was obtained:

Rainfall: 3   3   4   4   4.5 4.5 5   5   6   6   7   7
Maize: 54 61 65 68 66   70   68 78 83 81 90 94

(a) Plot the data in the form of a scatter diagram. Does there appear to be a relationship between the July rainfall and the yield of maize?
(b) Compute the coefficient of correlation and determination and interpret your answers.
(c) Compute the rank-order correlation coefficient.
(d) Compute the regression equation and estimate the yield of maize for a rainfall of 8 inches within a confidence level of 95 %.

Exercise 2 The management of an insurance company is interested in learning more about the variation in the number of car accidents. They would like to determine what relationship there is between temperature and the frequency of accidents. The following data have been collected:

Daily minimum temperature (°C): –1   8   24 16 19 12 15 –6
No of car accidents: 4   5   10   8   9   6   7   3

(a) Plot a scatter diagram and draw the line of best fit.
(b) Estimate, with a confidence level of 90 %, the number of accidents when the temperature is 8.5 °C.
(c) Compute and interpret the the coefficients of correlation and determination.
(d) Test the hypothesis that the coefficient of correlation in the population is zero. Use $\alpha = 0.05$

Exercise 3 The following data reflects the family income and food expenditure (in R'000) of a sample of ten families:

Income: 24   15   18.2 12.6 8   9.5 21   11.4 6.4 13.2
Expenditure: 3.6 2.9 2.9   2.6   2   2.4 3   2.5   1.8 2.4

(a) Plot a scatter diagram.
(b) Determine the regression equation.

(c) Estimate, within a confidence level of 99 %, the food expenditure of a family with an income of R20 000.
(d) Determine and interpret the coefficient of correlation and the coefficient of determination.
(e) Test the reliability of the correlation coefficient.
(f) Determine the coefficient of rank correlation.

Exercise 4 The head of the local traffic department is concerned about the increase in traffic accidents in recent years. He is convinced that the increased number of taverns has a direct relationship to the accident rate. He has compiled the following information over the past nine years:

| No. of accidents: | 18 | 14 | 17 | 22 | 27 | 31 | 23 | 26 | 30 |
| No. of taverns: | 6 | 6 | 7 | 8 | 9 | 9 | 8 | 9 | 10 |

(a) Develop a regression equation and line of best fit for the data.
(b) Compute the sample coefficient of correlation and determination.
(c) Test the significance of the correlation coefficient. ($\alpha = 0.05$)
(d) Last month two taverns were closed down because of selling drinks after 1 am, so in the coming month there will be only 7 taverns in operation. Compute the expected number of accidents for the coming month within a confidence interval of 95 %.
(e) Compute Spearman's Rank-order correlation coefficient.

Exercise 5 The following data shows the present maintenance costs and age of a sample of 8 similar machines used in a factory:

Machine:	A	B	C	D	E	F	G	H
Age:	1	3	4	4	6	7	7	8
Maintenance (R):	200	550	650	800	1150	1100	1300	1500

(a) Calculate and plot a regression line of machine cost on machine age.
(b) Determine the coefficient of Rank correlation.

Exercise 6 The following table sets out the advertising costs and net profit of ten firms as a percentage of total sales:

Firm	Advertising (% of turnover)	Net Profit (% of turnover)
Rave	9	22
McInnes Co	6	19
Hamo	4.5	20
Mag Wheels	5	16
Genkon	8.5	26
Grant's	4	10

SIMPLE REGRESSION AND CORRELATION

Firm	Advertising (% of turnover)	Net Profit (% of turnover)
Kate Cake	3	12
Fat is Fun	5	14
Armstrong	3	9
Toys Galore	8	33

(a) Plot the scatter diagram and the line of best fit.
(b) Estimate profit as a percentage of turnover if advertising cost is 25 % of turnover.
(c) Estimate a 98 % confidence interval for net profit as a percentage of turnover when advertising costs are 7 % of turnover.
(d) Calculate the correlation coefficient and determine its reliability at a 2 % level of significance.

Exercise 7 Ten sales agents of a firm had the following numbers of years of service.

Agent:	A	B	C	D	E	F	G	H	I	J
Years:	8	6	4	12	5	3	1	14	9	10

The manager of the company arranged the agents in the following order of excellence: (most excellent to least excellent)
H I D J A E C B G F
Determine Spearman's rank-order correlation coefficient between years of service and excellency.

Exercise 8 Given the following data, compute a regression line and all relevant diagnostic statistics, and make all desirable interpretations. Use the step-by-step procedure and $\alpha = 0.05$ as guideline.

Output (y):	28	25	27	15	13	10	9	20
Hours worked:	7.3	6.1	5.7	5.0	5.2	4.4	3.9	12.0

Exercise 9 Opinions on a relationship between aptitude in mathematics and music are diverse. The following table lists the results for 17 students:

Maths mark (%): 56 59 62 63 68 69 72 73 74 76 79 82 85 87 94 96 97
Music mark (%): 48 51 58 78 62 66 74 76 71 80 81 83 72 90 92 93 96

(a) Plot a scatter diagram.
(b) Calculate the line of best fit using the least squares method and use the line to estimate the music result for a student with a maths mark of 80 %

(c) Calculate the coefficients of correlation and determination and comment on the reliability of your estimate.

Exercise 10 From the following data, determine the resting pulse rate that you would expect from someone exercising for a daily average of:
(a) 45 min; (b) 15min; and (c) 2.5 h

Daily exercise (min):	20	30	60	10	100	0	120	160	160	180
Pulse/min:	75	70	70	85	50	90	60	52	48	64

CHAPTER 12
Time Series Analysis

CONTENTS

12.1 INTRODUCTION 201
12.2 THE CLASSICAL MULTIPLICATIVE MODEL 201
12.3 HISTORIGRAM 203
12.4 ANALYSIS OF TIME SERIES 203

12.1 INTRODUCTION

Time series is a study of the behaviour of data collected over a period of time and arranged in a chronological order. Our purpose will be to see how the data observed change over time, and to try and predict what the future behaviour of that data will be. Thus, time series analysis helps us cope with uncertainty about the future.

Forecasting is an essential tool in any decision-making process. The quality of the forecasts management can make, is strongly related to the information that can be extracted and used from past data.

The basic assumption underlying time-series analysis is that those factors which have influenced patterns of economic activity in the past and present will continue to do so in more or less the same manner in the future. Thus the major goals of time-series analysis are to identify and isolate these influencing factors for forecasting purposes as well as for managerial planning and control. To achieve these goals, the most popular model is the classical muliplicative model.

12.2 THE CLASSICAL MULTIPLICATIVE MODEL

This model attempts to explain the pattern observed in an actual time series ($\hat{y}$) by the presence of four components:

(1) Secular trend (T)
(2) Cyclical fluctuation (C)
(3) Seasonal variation (S)
(4) Irregular variation (I)

The first two components can be classified as the long-term trend and the second two as short-term trend.

In equation form the relationship between components and the original time series can be described as: $\hat{y} = T \times C \times S \times I$

Secular trend This is the general increase or decrease in the time series values over an extended number of years. The steady increase in the cost of living recorded by the Consumer Price Index is an example of secular trend. From year to year the cost of living varies a great deal, but if we examine a long-term period, we see that the trend is towards a steady increase.

Cyclical movement These variations refer to long-term swings about a general trend line. The most common example of cyclical fluctuation is the business cycle of prosperity, recession, depression and recovery. These movements don't follow any definite trend but move in a somewhat unpredictable manner.

Seasonal variation These variations are patterns that occur within a period of one year and tend to be repeated from year to year. For example a doctor can expect a substantial increase in the number of flu cases every winter. Since these are regular patterns, they are useful in forecasting the future.

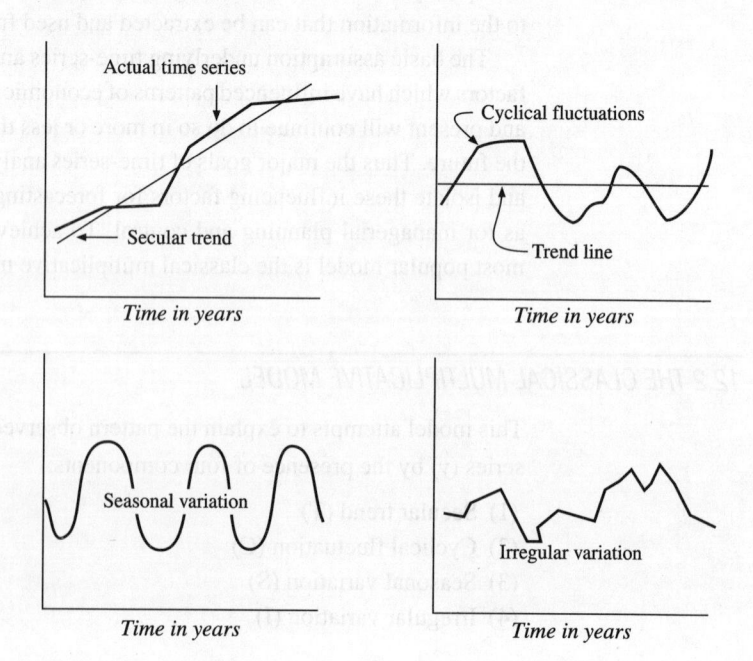

TIME SERIES ANALYSIS

Irregular variation These random variations occur over short intervals. The variations are unpredictable and there is no pattern to their behaviour, both as to when they will occur and as to their duration. Examples of these movements are floods, strikes, wars, etc. Because of their unpredictability, we do not attempt to explain them mathematically.

The irregular component is what is left after the other components have been estimated.

12.3 HISTORIGRAM

A graph constructed by making use of historical information over a time period is known as a historigram.

(1) Time is the independent variable and is measured along the horizontal axis. The amounts or values are the dependent variables and are measured on the vertical axis.

(2) The points plotted to represent the values against time are linked by straight lines (line graph).

12.4 ANALYSIS OF TIME SERIES

Secular trend analysis In trend analysis, two decisions must be made:

(1) What shape line will be used to represent the relationship between the two variables? This book will deal only with

Straight line: $\hat{y} = a + bx$

The procedure is to plot a scatter diagram from which a judgement can be made.

(2) What method will be used to determine the coefficients of an equation that will describe the relationship?
— Method of least squares
— Method of semi-averages
— Method of moving averages

Method of least squares This criterion requires that the sum of the squares of the differences between the actual y values and the calculated trend values ($\hat{y}$) be a minimum.

$$\sum (y - \hat{y})^2 = \text{minimum}$$

Step-by-step procedure:

❑ The y values are usually recorded annually over several consecutive years.
❑ The time variable (x) can be coded to simplify calculations.
  Coding:
  — Find the mean time or origin. That is the middle of the time period.
  — Subtract the mean time from each of the sample times.
  — When there is an odd number of years, the origin is at the middle of the year (1 July), and when there is an even number of years, the origin is at the end of the year (31 December)
  — The code at the origin is zero.
❑ Determine the trend equation:

$$\hat{y} = a + bx$$

where:

$$a = \frac{\sum y}{n}$$

$$b = \frac{\sum xy}{\sum x^2}$$

❑ Fit a straight line of the form $\hat{y} = a + bx$ to the data and read off the forecast value from the graph on the y-axis.
❑ If you use the trend equation to forecast, first determine an appropriate code value for x (the year of forecast) and substitute it in the formula.
❑ Modifications to the trend equation:
  — modify a yearly trend equation so that it can be used with monthly or quarterly data.
  — modify the equation so that the origin is in the middle of a month or quarter instead of in the middle or end of a year.

Example:

Year	x	y	x^2	xy
1975–1978	= –3	80	9	–240
1976–1978	= –2	95	4	–190
1977–1978	= –1	100	1	–100
1978–1978	= 0	110	0	0
1979–1978	= 1	130	1	130
1980–1978	= 2	145	4	290
1981–1978	= 3	150	9	450
		810	28	340

$$a = \frac{\sum y}{n}$$

$$= \frac{810}{7}$$

$$= 115.71$$

$$b = \frac{\sum xy}{\sum x^2}$$

$$= \frac{340}{28}$$

$$= 12.14$$

$\hat{y} = a + bx$

$\quad = 115.71 + 12.14x$

Origin: 1 July 1978
x unit: 1 year
y unit: annual sales

Forecast for 1982:
$\hat{y} = 115.71 + 12.14(4)$
$\quad = 164.27$

Monthly equation:
$$\hat{y} = \frac{115.71}{12} + \frac{12.14}{12} \cdot \frac{x}{12}$$

$\quad = 9.64 + 0.08x$

Move origin from 1 July 1978 to the middle of January 1982.

Annual Sales

$\hat{y} = 115.71 + 12.14x$

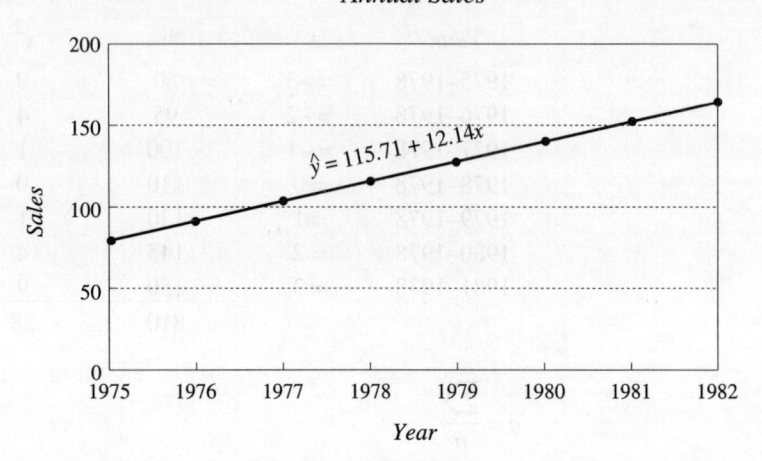

There are 42.5 trend increments of 0.08 each which must be added to the constant (a):

$\hat{y} = 9.64 + 0.08x$

$= 9.64 + 42.5(0.08) + 0.08x$

$= 13.04 + 0.08x$

x unit: 1 month
y unit: monthly sales
origin: 15 January 1982

This information can be used to make a monthly forecast for 1982:
January 1982: $13.04 + 0.08(0) = 13.04$
February 1982: $13.04 + 0.08(1) = 13.12$

Method of semi-averages

- Split the data down the middle into two equal parts. If there is an odd number of years, simply omit the middle year.
- Calculate the arithmetic mean for each part.
- Plot the two means at the midpoint of the time intervals covered by the respective parts.
- Join these two points with a straight line — this is the trend line.
- To forecast we can use the graph by just reading the value from the graph on the *y* axis; or
- calculate the average increase per year and then add this increment an appropriate number of times to a trend value that is already known (one of the group averages).

TIME SERIES ANALYSIS

Example

Year	Sales (R)	
1975	80	
1976	95	$=\dfrac{275}{3}=92$
1977	100	
1978	110	
1979	130	
1980	145	$=\dfrac{425}{3}=142$
1981	150	

Trend value for 1 July 1976 = 92
Trend value for 1 July 1980 = 142

Sales 1975–1982

[graph of sales from 1975 to 1982 showing linear trend]

Average increase per year $= \dfrac{142-92}{4} = 12.5$

Forecast for 1982: $142 + 2(12.5) = 167$

Method of moving averages (smoothing of time series)   When we are not interested in obtaining a mathematical equation, this method is useful. It is an artificially constructed time series in which each annual figure is replaced by the average of itself and values to a number of preceding and succeeding periods. This period can be any number of years, but longer periods make smoother curves. If the period is too long, a straight line will result and the general direction of the curve will be lost.

The moving average reduces the fluctuations of a series and gives it a much smoother appearance, but some values are lost at each end of the series.

This method is strongly affected by extreme values.

Example:

Year	y	3-year moving average	4-year moving average	Centred 4-year average
1977	2			
1978	6	$(2 + 6 + 1) \div 3 = 3$	$(2 + 6 + 1 + 5) \div 4 = 3.5$	
1979	1	$(6 + 1 + 5) \div 3 = 4$	$(6 + 1 + 5 + 3) \div 4 = 3.75$	3.62
1980	5	$(1 + 5 + 3) \div 3 = 3$	$(1 + 5 + 3 + 7) \div 4 = 4.0$	3.88
1981	3	$(5 + 3 + 7) \div 3 = 5$	$(5 + 3 + 7 + 2) \div 4 = 4.25$	4.12
1982	7	$(3 + 7 + 2) \div 3 = 4$		
1983	2			

Smoothing of time series

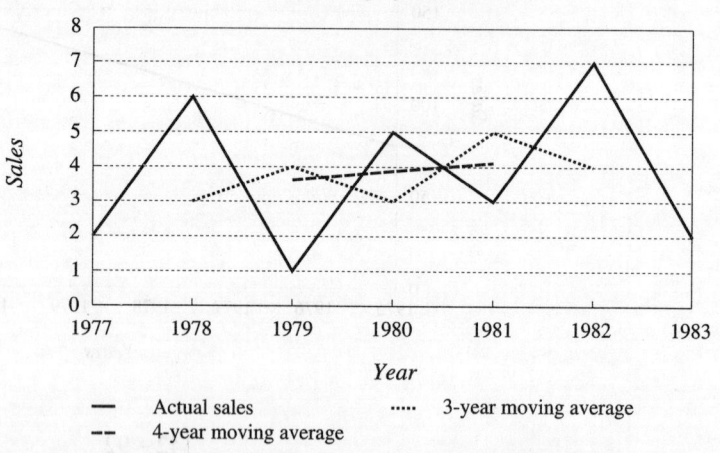

― Actual sales ····· 3-year moving average
-- 4-year moving average

Seasonal variation Seasonal variations occur within a period of 1 year or less, therefore period data is required (weekly, monthly, quarterly, daily). Seasonal variation is generally expressed as an index number and this index number can be obtained by making use of the method of ratio-to-moving-average.

Ratio-to-moving-average-method (Percentage of moving average method)

❏ List data in chronological order.

- Calculate the required moving average.
- If the period is an even number, centre the averages by averaging adjacent moving averages.
- Compute the seasonal values by dividing the actual data by the moving averages and multiply by 100.
- Construct a summary table of the seasonal percentages and by making use of the modified mean approach, compute an unadjusted seasonal index. A modified mean is the arithmetic mean of the values that remain after elimination of the smallest and largest values in the column.
- Determine the factor needed to adjust index numbers to a typical index number.

    Typical index: Quarterly = $100 \times 4 = 400$
    Monthly $\qquad = 100 \times 12 = 1\,200$

    $$\text{Factor} = \frac{\text{Total typical index}}{\text{Total of unadjusted indexes}}$$

- Calculate the adjusted typical seasonal indices by multiplying unadjusted indices with the factor.
- Deseasonalizing or seasonally adjusting data is the adjustment of data to show how things would have been had there been no seasonal fluctuations. This process consists of dividing each value in a series by the appropriate typical seasonal index for that period and multiplying the result by 100.

Seasonalized forecasting

The obvious way to forecast trend of time series is to extrapolate from the trend equation describing the historical data. Extrapolation means extending the trend into the future to estimate a value that lies beyond the range of the values from which the trend equation was derived.

Steps:

- Determine the long-term trend equation by making use of the method of least squares.
- Modify the trend equation for use with the seasonal period (monthly, quarterly, etc.).
- Shift the origin to the middle of the first forecast period.
- Determine the code for the forecast period and substitute it in place of x in the equation.
- The result is a deseasonalized forecast value.

❑ To seasonalize the forecast, multiply each forecasted value with the typical index for that period.

Example

Year	Quarter	Sales	4-quarter moving average	Centred average	Percentage (2 ÷ 4)
	(1)	(2)	(3)	(4)	
1970	I	8			
	II	12			
	III	12	10.5	10.75	111.63
	IV	10	11.0	11.38	87.87
1971	I	10	11.75	12.0	83.33
	II	15	12.25	12.25	122.45
	III	14	12.25	12.12	115.51
	IV	10	12.0	11.88	84.18
1972	I	9	11.75	11.75	76.6
	II	14	11.75	12.0	116.67
	III	14	12.25	12.38	113.09
	IV	12	12.5	12.75	94.12
1973	I	10	13.0	13.38	74.74
	II	16	13.75	13.88	115.27
	III	17	14.0		
	IV	13			

Summary table to compute adjusted typical indices

YEAR	I	II	III	IV	TOTAL
1970			111.63	87.87	
1971	83.33	122.45	115.51	84.18	
1972	76.60	116.67	113.09	94.12	
1973	74.74	115.27			

Modified unadjusted index

76.60   116.67   113.09   87.87 = 394.23

Factor: $= \dfrac{400}{394.23} = 1.0146$

TIME SERIES ANALYSIS

Typical index:
77.72   118.38   114.74   89.16   = 400

Deseasonalized sales for 1974

Year	Quarter	Sales	Seasonal Index	Deseasonalized Sales
1971	I	10	77.72	10 ÷ 0.7772 = 12.87
	II	15	118.38	15 ÷ 1.1838 = 12.67
	III	14	114.74	14 ÷ 1.1474 = 12.20
	IV	10	89.16	10 ÷ 0.8916 = 11.22

The example shows an increase in actual sales from the first to the second quarter, but deseasonalization data changes this to a decrease.

Forecast for 1974 using the trend and seasonal components

Year	Yearly sales (y)	x	x^2	xy
1970	42	−1.5	2.25	−63.0
1971	49	−0.5	0.25	−24.5
1972	49	0.5	0.25	24.5
1973	56	1.5	2.25	84.0
	196	0.0	5.00	21.0

$$a = \frac{\sum y}{n}$$
$$= \frac{196}{4}$$
$$= 49$$

$$b = \frac{\sum xy}{\sum x^2}$$
$$= \frac{21}{5.0}$$
$$= 4.2$$

$\hat{y} = 49 + 4.2x$ with x unit = 1 year

Origin: 31 December 1971

Equation for quarterly data:

$$\hat{y} = \frac{49}{4} + \frac{4.2}{4}\left(\frac{x}{4}\right)$$

$$= 12.25 + 0.26x \text{ with } x \text{ unit} = 1 \text{ quarter}$$

Origin: 31 December 1971

Move origin to middle of first quarter 1974
$$\hat{y} = 12.25 + 8.5(0.26) + 0.26x$$

$$= 14.46 + 0.26x$$

Forecast:

Year	Quarter	x	$\hat{y}$	Typical index	Seasonalized Forecast
1974	I	0	14.46	77.72	11.24
	II	1	14.72	118.38	17.43
	III	2	14.98	114.74	17.19
	IV	3	15.24	89.16	13.59

Cyclical variation The cyclical variation is the long-term movement related to the business cycle. The method typically used for estimating the size of this component is known as the residual method. This method involves the removal of both the seasonal and trend factors from the composite cycle ($\hat{y} = T \times S \times C \times I$) and the conservation of the fluctuations which remain.

(Use the example on page 210)

Step 1   Construct a table with columns 1 and 2, the given time periods and sales respectively.

Step 2   Calculate a trend equation for the data and the trend values (T).
$\hat{y} = 14.46 + 0.26x$ with x-unit = 1 quarter and origin: middle of 1st quarter 1974

Step 3   Calculate the typical seasonal indices (S) per quarter and write it down in column 4.

Step 4   Multiply each T × S in column 5.

Step 5   Divide sales (y) by T × S. That is column 2 ÷ column 5 × 100. The result reflects irregular and cyclical variations.

Step 6 The irregularities can be smoothed out by making use of a 3-period weighted moving average. This weighted average is computed by using weights of 1 : 2 : 1. The first quarter is multiplied by 1, the second quarter by 2 and the third quarter by 1. These values are then summed and divided by 4 (sum of weights: 1 + 2 + 1 = 4) in order to provide a cyclical component.

Table

1		2	3	4	5	6	7
Year		Sales	x-code	T	S ÷ 100	T × S	2 ÷ 5 × 100	C
1970	I	8	−16	10.3	0.7772	8.005	99.94	
	II	12	−15	10.56	1.1838	12.501	95.99	97.14
	III	12	−14	10.82	1.1474	12.415	96.66	97.63
	IV	10	−13	11.08	0.8916	9.879	101.22	103.14
1971	I	10	−12	11.34	0.7772	8.813	113.47	109.35
	II	15	−11	11.60	1.1838	13.732	109.23	108.70
	III	14	−10	11.86	1.1474	13.608	102.88	101.88
	IV	10	−9	12.12	0.8916	10.806	92.54	95.38
1972	I	9	−8	12.38	0.7772	9.622	93.54	93.30
	II	14	−7	12.64	1.1838	14.963	93.56	93.81
	III	14	−6	12.90	1.1474	14.801	94.59	96.26
	IV	12	−5	13.16	0.8916	11.733	102.28	98.76
1973	I	10	−4	13.42	0.7772	10.430	95.88	98.21
	II	16	−3	13.68	1.1838	16.194	98.80	99.94
	III	17	−2	13.94	1.1474	15.995	106.28	103.51
	IV	13	−1	14.20	0.8916	12.661	102.68	
1974	I		0					

Analysis of cyclical influence is useful for historical analysis, but its value in forecasting is very limited. Instead, a number of business indicators are used to forecast cyclical turning points.

EXERCISES

Exercise 1   The following data are population data:

Year	Population (in millions)
1970	204.9
1971	207.1
1972	208.8
1973	210.4
1974	211.9
1975	213.6
1976	215.2
1977	216.9
1978	218.5
1979	220.6

(a) Plot the data on a graph.
(b) Fit a straight line trend equation to the data using the least squares method.
(c) Estimate the 1981 population.

Exercise 2   The following data represents the maize production for a particular farm.

Year	Tons
1976	200
1977	205
1978	215
1979	210
1980	230
1981	240
1982	245

(a) Plot the data on a graph.
(b) Fit a trend equation to the data using the least squares method.
(c) Estimate the 1983 production.

TIME SERIES ANALYSIS

Exercise 3 The following data are recent income security payments by the government:

Fiscal Year	Payments (billions)
1977	73.0
1978	84.4
1979	108.6
1980	127.4
1981	137.9
1982	146.2
1983	160.2
1984	193.1
1985	225.1

(a) Estimate the 1986 expenditure by making use of the method of least squares.

(b) Estimate the 1986 expenditure by making use of the method of semi-averages.

Exercise 4 The sales of two organizations are given below.

Year	Sales of I ($\times R1\ 000$)	Sales of II ($\times R1\ 000$)
1977	373	430
1978	388	438
1979	398	447
1980	417	453
1981	430	462
1982	445	470

(a) Use the method of least squares and fit a linear equation to each set of data.

(b) Assume that the current trend will continue, and determine the year in which the sales of I will first exceed the sales of II.

Exercise 5 The president of the National Motor Company is studying his compact car sales over the last 5 years:

Year:	1979	1980	1981	1982	1983
Number sold:	794	865	931	1041	1150

(a) Use the method of least squares to estimate National's 1984 sales.

(b) Use the method of semi-averages to estimate the 1984 sales.

Exercise 6 The owner of Progress Builders is examining the number of solar homes on which construction was started in the region in each of the last seven months.

Month:	June	July	Aug	Sept	Oct	Nov	Dec
Number of homes:	15	15	26	27	33	41	51

(a) Plot these data.
(b) Develop the linear estimating equation that best describes these data and plot the curve on the graph.
(c) Estimate sales for March next year using both the graph and the equation.

Exercise 7 The Tasty Hamburger chain has significantly increased its investments in inventory over the last six years. The information is as follows:

Year:	1977	1978	1979	1980	1981	1982
Inventory ($\times R10\ 000$):	4	4.5	6	8	8.5	10

(a) Plot these data.
(b) Develop the linear estimating equation that best describes these data, and plot the graph.
(c) Estimate the 1984 inventory.

Exercise 8 Use the data given in the following table to smooth the time series by making use of:
(a) a 2-year moving average;
(b) a 3-year moving average; and
(c) a 4-year moving average.

1975	1976	1977	1978	1979	1980	1981	1982	1983
642	819	845	755	767	720	749	749	686

Exercise 9 A company's sales for the years 1971 to 1979 were as follows: ($\times$ R1 000)

1971	1972	1973	1974	1975	1976	1977	1978	1979
324	296	310	305	295	347	348	364	370

(a) Derive, by the method of least squares, an equation of linear trend for the sales of the company.
(b) Compute trend values for the years 1980 and 1981.
(c) Sketch the graph and the trend line of the series.

(d) Draw a trend line using the method of semi-averages and make estimates for 1980 and 1981.

(e) Draw a trend line using the 3-year moving average method and make estimates for 1980 and 1981.

Exercise 10 Use the following trend equation to answer the questions below:
$$\hat{y} = 284 + 14.4x \text{ with origin: 1 July 1974}$$
$$x \text{ unit: 1 year}$$
$$y \text{ unit: annual sales } (\times \text{R1 000})$$

(a) If 1982 was the last year of the data, how many years of data were used to determine the trend equation assuming the origin has not been moved?

(b) Project sales for 1985.

(c) What is the annual rand increase in sales?

(d) Convert the equation to a monthly equation with the x unit = 1 month and set the origin at 15 July 1980. Estimate sales for September 1980.

Exercise 11 Use the following trend equation to answer the questions below.
$$\hat{y} = 50 + 12x \text{ with origin: 31 December 1974}$$
$$x \text{ unit: 6 months}$$
$$y \text{ unit: annual production}$$

(a) If 1982 was the last year of data, how many years of data were used to determine the trend equation if the origin has not been changed?

(b) What is the annual rand increase in sales?

(c) Project sales for 1983.

Exercise 12 Use the following equation to answer the questions below.
$$\hat{y} = 100 + 10x \text{ with origin: 31 December 1978}$$
$$x \text{ unit: 6 months}$$
$$y \text{ unit: annual production}$$

(a) Estimate production for 1983.

(b) Convert the equation to a monthly equation with x unit = 1 month.

(c) Shift origin to 15 July 1982.

(d) Estimate production for December 1982.

Exercise 13 The owner of the Pleasure Glide Boat Company has compiled the following quarterly figures (R'000) of the company's investment in accounts receivable over the last 5 years.

Year	Spring	Summer	Autumn	Winter
1979	101	118	90	79
1980	108	123	94	83
1981	109	125	96	86
1982	113	131	102	91
1983	119	140	108	97

(a) Calculate quarterly seasonal indices by making use of the ratio-to-moving-average method.

(b) Deseasonalize the 1981 data.

(c) Estimate the seasonalized investment for winter 1985.

Exercise 14 Wheeler Airline has estimated the number of passengers for December at 595 000 (deseasonalized). How many passengers should they anticipate if the December seasonal index is 128?

Exercise 15 A Sea Fisheries research group has measured the level of mercury contamination in the ocean at a certain point off the East Coast. The following percentages of mercury were found in the water:

	Jan	Feb	Mar	Apr	May	Jun	Jul	Aug	Sep	Oct	Nov	Dec
1981	0.4	0.6	0.9	0.7	0.8	0.6	0.7	0.5	0.5	0.6	0.3	0.4
1982	0.5	0.8	0.8	0.9	0.6	0.7	0.8	0.6	0.5	0.5	0.4	0.3
1983	0.3	0.5	0.7	0.8	0.8	0.6	0.9	0.7	0.6	0.5	0.4	0.4

Construct monthly typical seasonal indices for the data.

Exercise 16 The production manager of a paper mill has accumulated the following information on quarterly production (in millions of rands):

	Winter	Spring	Summer	Autumn
1980	2.6	4.1	4.8	3.2
1981	2.9	4.5	5.0	3.4
1982	2.8	4.9	5.5	3.3
1983	3.1	5.1	5.6	3.6

(a) Calculate the typical seasonal indices.

(b) Deseasonalize these data.

(c) Find the least squares line that best describes these data.

(d) Estimate the value of the production during the spring of 1984.

TIME SERIES ANALYSIS

Exercise 17 Sales of Glady Plastic were R84 000 for the first quarter of 1982.

(a) If the seasonal indices per quarter are: 75; 90; 125 and 110, what are the deseasonalized sales for the first quarter?

(b) Assuming average quarterly sales of R100 000, estimate the sales for the third quarter.

Exercise 18 A large Receiver of Revenue has determined the following seasonal indices for sales tax collections in a certain area:

Quarter:	I	II	III	IV
Index:	70	120	80	130

If the annual sales tax collections are estimated to be R1.2 billion, how much should be collected in the second quarter of the year?

Exercise 19 The managing director of a skin-diving equipment wholesale house assembled the following information regarding his quarterly revenue (× R10 000)

	I	II	III	IV
1980	70.2	58.4	52.7	77.9
1981	70.4	59.8	54.2	80.3
1982	72.1	57.5	52.0	83.5
1983	73.6	63.5	57.9	88.2

(a) Calculate the seasonal indices for these data.
(b) Deseasonalize the data for 1983.
(c) Find the least squares line that best describes these data.
(d) Estimate the revenue for the first quarter of 1984.

Exercise 20 An official in the Department of Trade and Industry has the following data describing the value of wheat exported from a certain region during the last 4 years.

	I	II	III	IV
1981	1	3	6	4
1982	2	2	7	5
1983	2	4	8	5
1984	1	3	8	6

(a) Determine the seasonal indices and deseasonalize these data.
(b) Calculate the least squares line that best describes these data.
(c) Plot the original data, the deseasonalized data and the trend line.

CHAPTER 13
Index Numbers

CONTENTS

13.1 INTRODUCTION 221
13.2 PROBLEMS IN CONSTRUCTION
OF INDEX NUMBERS 222
13.3 CONSTRUCTION OF A SIMPLE INDEX NUMBER . . 222
13.4 CONSTRUCTION OF A COMPOSITE OR
AGGREGATE INDEX NUMBER 223
13.5 CHANGING OF THE BASE PERIOD 229
13.6 THE CONSUMER PRICE INDEX
OF SOUTH AFRICA 221

13.1 INTRODUCTION

An index number is a relative figure expressed as a percentage which is used to measure how much an economic variable changes over time or differs between two locations. We calculate an index number by finding the ratio of the current value (numerator) to a base value (denominator) and then multiply the resulting number by 100. This final value is the percentage relative.

Measurements compared by an index number can be concerned with price, quantity or value. A price index is the one most frequently used. It compares changes in price from one period to another. A quantity index measures how much the number or quantity of a variable changes over time. The value index measures changes in total monetary worth. It combines price and quantity changes to present a more informative index.

Some examples of index numbers are:
❏ The consumer price index
❏ The wholesale price index
❏ Labour indices on employment
❏ New car sales index
❏ Stock exchange index

13.2 PROBLEMS IN CONSTRUCTION OF INDEX NUMBERS

(1) What is the purpose of the index?
(2) Availability and selection of suitable data related to the purpose. Because index numbers are generally based on a sample, items must be relevant and representative of the general situation.

For example: Figures reported on an annual basis are not suitable to compute an index describing seasonal variation. The Producers price index should be composed of wholesale prices while a cost of living index should be based on retail prices.
(3) Comparability of data: Data must have comparable characteristics and be of the same quality. It is essential for the data to be as accurate as possible.
(4) Choice of the base period: The period relative to which the comparison is made, is called the base period.
 - The period should be recent enough so that comparisons with the base are meaningful.
 - It should also be a typical period with respect to the activity of interest.
 - The period should be one of relative economic stability.
 - The index number for the base year is always 100.
(5) Selection of a weighting system: All the items included in an index number are usually not of equal importance, therefore it is necessary to apply a system of weights so that the index will reflect the relative importance of its components.
(6) Determining method for computing the index:
Index numbers are differentiated according to the number of commodities included in the comparison.
 - Simple index number: this represents a comparison for a single product or commodity.
 - Aggregate or composite index number: these indices are constructed for groups of products or commodities.

13.3 CONSTRUCTION OF A SIMPLE INDEX NUMBER

(1) Select the value to be used as base.
(2) Divide every value in the series by the base value.
(3) Multiply the ratio of each value in the series to a particular value in the series by 100.

INDEX NUMBERS

$$Ip \text{ (simple price index)} = \frac{Pi}{Pb} \times 100$$

$$Iq \text{ (simple quantity index)} = \frac{Qi}{Qb} \times 100$$

Where:

Pi = price of commodity in given period
Pb = price in the base period
Qi = quantity of item in given period
Qb = quantity in base period

These simple index numbers are called relatives.

13.4 CONSTRUCTION OF A COMPOSITE OR AGGREGATE INDEX NUMBER

An index number can be constructed for a group or series of items where an overall change in figures, rather than individual changes, is being investigated.

Classification of methods

(1) Unweighted aggregate index
(2) Unweighted average of relatives index
(3) Weighted aggregate index
(4) Weighted average of relatives index

In the aggregative type of index, variables are summed for all the items in the base period as well as the current period, and the index is simply the ratio of these two aggregates.

In the average of relatives type of index, a simple index is first computed for each commodity using current and base values. These relatives are then summed and the average is determined.

Unweighted (simple) aggregate index This index should only be used in cases when all items are reported in the same units and all items are of equal importance.

❏ State the formula.
❏ Sum the columns for each year.
❏ Evaluate the formula for each period.

$$Ip = \frac{\sum Pi}{\sum Pb} \times 100$$

$$Iq = \frac{\sum Qi}{\sum Qb} \times 100$$

Unweighted average of relatives index By converting all figures to relatives and finding their averages, the effect of the measurement unit is eliminated. All items in the index have to be weighted equally.
- State the formula.
- Express each number in the table as a simple index or relative.
- Sum the simple indices for each year.
- Multiply the answer by 100.
- Divide the result by the number of commodities (n).

$$Ip = \frac{\sum \frac{Pi}{Pb}}{n} \times 100$$

$$Iq = \frac{\sum \frac{Qi}{Qb}}{n} \times 100$$

Weighted aggregate index The most common type of weight used in constructing an index, is quantity. This assumes that the quantity of a commodity in a given period is a valid measure of its comparative importance. A weighting factor is added, so the index will reflect the relative importance of its components.

Laspeyres index When an index number is constructed using base year quantities as weights, it is known as Laspeyres Index. This index measures the change in the total cost of a fixed quantity of goods that represent a consumption pattern typical of the base period. The problem is that consumption patterns change with time and when this pattern changes, the base period should also be changed. A feature is that it requires quantity information for only one period. This index can generally be expected to overestimate the change in total cost (upward bias).

INDEX NUMBERS

$$Ip(L) = \frac{\sum PiQb}{\sum PbQb} \times 100$$

$$Iq(L) = \frac{\sum QiPb}{\sum QbPb} \times 100$$

Paasche's Index This index uses current year quantities as weights. It measures the change in the total cost of a fixed quantity of goods that represent a consumption pattern typical of the current year. The use of current year weights avoids the problem of changing consumption patterns, but requires that the weighting system be continually re-established (which is expensive and difficult). This index shows a downward bias (that is, it tends to under-estimate the change in total cost).

$$Ip(P) = \frac{\sum PiQi}{\sum PbQi} \times 100$$

$$Iq(P) = \frac{\sum QiPi}{\sum QbPi} \times 100$$

In general the difference between the Paasche's and Laspeyres indices is small. Both are weighted indices that give consideration to the comparative importance of the various quantities included in the index and also eliminate the effect of the different measurement units that might have been used.

Irving Fisher's ideal index For this index the indices of Laspeyres and Paasche are averaged by making use of the Geometric Mean.

$$Ip(IF) = \sqrt{Ip(P) \cdot Ip(L)}$$

$$Iq(IF) = \sqrt{Iq(P) \cdot Iq(L)}$$

Drobisch's index For this index the indices of Laspeyres and Paasche are averaged by making use of the arithmetic mean.

$$Ip(D) = \frac{Ip(L) + Ip(P)}{2}$$

$$Iq(D) = \frac{Iq(L) + Iq(P)}{2}$$

Example Prices and consumption of three commodities 1980–1986

Commodity	1980 (Pb)	1986 (Pi)	1980 (Qb)	1986 (Qi)
Milk	0.30	0.38	30	35
Bread	0.25	0.35	3.8	3.7
Eggs	0.60	0.90	1.5	1.0
Total	1.15	1.63	35.3	39.7

$\frac{Pi}{Pb}$	$\frac{Qi}{Qb}$	PiQb	PbQb	QiPb	PiQi
1.27	1.17	11.40	9.0	10.50	13.3
1.40	0.97	1.33	0.95	0.92	1.3
1.50	0.67	1.35	0.90	0.60	0.9
4.17	2.81	14.08	10.85	12.02	15.50

(a) Simple price/quantity indices for 1986 with 1980 = 100

Milk: $Ip = \frac{Pi}{Pb} \times 100$

$= \frac{0.38}{0.30} \times 100$

$= 126.7$

Increase of 26.7 % in the price of milk between 1980 and 1986.

$Iq = \frac{Qi}{Qb} \times 100$

$= \frac{35}{30} \times 100$

$= 116.7$

Increase of 16.7 % in the quantity of milk sold between 1980 and 1986.

Eggs: $Ip = \frac{0.90}{0.60} \times 100$

$= 150$

Increase of 50 % in the price of eggs between 1980 and 1986.

$Iq = \frac{1.0}{1.5} \times 100$

$= 66.7$

Decrease of 33.3 % in the quantity of eggs sold between 1980 and 1986.

(b) *Unweighted (simple) aggregate index*

$$Ip = \frac{\sum Pi}{\sum Pb} \times 100$$

$$= \frac{1.63}{1.15} \times 100$$

$$= 141.7$$

Increase of 41.7 % in the total price of the commodities between 1980 and 1986.

$$Iq = \frac{\sum Qi}{\sum Qb} \times 100$$

$$= \frac{39.7}{35.3} \times 100$$

$$= 112.5$$

Increase of 12.5 % in the total quantity of the commodities sold between 1980 and 1986.

(c) *Unweighted average of relatives index*

$$Ip = \frac{\sum \frac{Pi}{Pb}}{n} \times 100$$

$$= \frac{4.17}{3} \times 100$$

$$= 139$$

Average increase of 39 % in the relative price of the commodities between 1980 and 1986.

$$Iq = \frac{\sum \frac{Qi}{Qb}}{n} \times 100$$

$$= \frac{2.81}{3} \times 100$$

$$= 93.7$$

Average decrease of 6.3 % in the relative quantities sold between 1980 and 1985.

(d) *Laspeyres Index*

$$Ip = \frac{\sum PiQb}{\sum PbQb} \times 100$$

$$= \frac{14.08}{10.85} \times 100$$
$$= 129.8$$

Increase of 29.8 % in the total cost of the commodities between 1980 and 1985.

$$Iq = \frac{\sum QiPb}{\sum QbPb} \times 100$$
$$= \frac{12.02}{10.85} \times 100$$
$$= 110.8$$

Increase of 10.8 % in the total quantities sold between 1980 and 1985.

(e) *Paasche's Index*

$$Ip = \frac{\sum PiQi}{\sum PbQi} \times 100$$
$$= \frac{15.50}{12.02} \times 100$$
$$= 129.0$$

Increase of 29 % in the total cost of the commodities between 1980 and 1985.

$$Iq = \frac{\sum QiPi}{\sum QbPi} \times 100$$
$$= \frac{15.50}{14.08} \times 100$$
$$= 110.1$$

Increase of 10.1 % in the total quantities sold between 1980 and 1985.

(f) *Irving Fisher*

$$Ip = \sqrt{129.8 \times 129}$$
$$= 129.4$$

Increase of 29.4 % in the total cost of the commodities between 1980 and 1985.

$$Iq = \sqrt{110.8 \times 110.1}$$
$$= 110.4$$

Increase of 10.4 % in the total quantities sold between 1980 and 1985.

INDEX NUMBERS

(g) *Drobish*

$$Ip = \frac{129.8 + 129}{2}$$

$$= 129.4$$

Increase of 29.4 % in the total cost of the commodities between 1980 and 1985.

$$Iq = \frac{110.8 + 110.1}{2}$$

$$= 110.45$$

Increase of 10.45 % in the total quantities sold between 1980 and 1985.

13.5 CHANGING OF THE BASE PERIOD

Most index numbers are subject to revision from time to time due to major changes in conditions, standards of living, new formulae, inclusion of new items or disappearance of old ones, new techniques, etc. When a major change is made, it is customary to begin again with a new base, but it is not necessary to recalculate the old index numbers on the new system.

When a base period is no longer considered normal or typical, or when a more recent base is required, it is necessary to change the base period. When two or more series with different base periods are to be compared, their base periods need to be changed so that the comparison can be based on the same period for both.

$$\text{New index number} = \frac{\text{Original (old) index number}}{\text{Old index of new base period}} \times 100$$

Example *Value Indices for Ford Co sales*

Year:	1980	1981	1982	1983	1984	1985
1982 = 100:	74.2	81.4	100.0	114.0	117.0	118.9
1984 = 100:	63.4	69.6	85.5	97.4	100.0	101.6

1980: $\dfrac{74.2}{117.0} \times 100 = 63.4$

1982: $\dfrac{100}{117.0} \times 100 = 85.5$

1984: $\dfrac{117}{117.0} \times 100 = 100$

1985: $\dfrac{118.9}{117.0} \times 100 = 101.6$

13.6 THE CONSUMER PRICE INDEX OF SOUTH AFRICA

This index is a very important economic indicator and is used to determine inflation rate.

Method of determining the CPI

The formula used is that of the weighted aggregate index — Laspeyres' view.

$$CPI = \frac{\sum PiQb}{\sum PbQb} \times 100$$

To determine the base year weight factor, at least 10 000 households were sampled out of different income groups and metropolitan areas. Following international practice, the base period used for the CPI must change at least once every 10 years, and it has recently changed from 1985 to 1990 with effect from January 1990. A monthly index for each consumer item for each one of the strata is determined by making use of the above formula and after that a combined CPI is calculated.

The formula for determining inflation rate is as follows:

$$\text{Inflation rate} = \frac{\text{CPI current year}}{\text{CPI corresponding month previous year}}$$

EXERCISES

Exercise 1   The Jacobs Lawn Care Company provides monthly lawn fertilizing as well as disease and bug control service to northern suburbs home owners. The prices which Jacobs paid for its chemicals for 1974, 1976 and 1978 are shown below:

Chemical	Price		
	1974	1976	1978
10–10 (per 3 kg)	0.52	0.56	0.64
5–10 (per 4 kg)	0.48	0.53	0.56
XL (per 1 kg)	1.07	1.19	1.27
XX (per 1.5 kg)	0.94	1.09	1.24
DD (per 5 kg)	2.17	2.32	2.79

(*a*) Using an unweighted aggregates method with 1976 as the base year, determine the aggregate price index for 1974 and 1978.

(b) Determine a simple price index with 1976 as base for:
   (i) 5–10;
   (ii) XL; and
   (iii) DD

(c) Construct an unweighted average of relatives index for 1976 and 1978 with 1974 as base year.

Exercise 2 The Valdo Art School has compiled the following information showing prices and quantities of the following supplies for 1983 to 1985.

Items	Prices			Quantities		
	1983	1984	1985	1983	1984	1985
Brushes	4.74	4.92	4.93	50	54	43
Canvas	3.10	3.41	3.94	920	907	878
Oil	6.29	6.83	7.00	107	121	137

(a) Determine the weighted aggregate price index for 1984 and 1985 using the Laspeyres method with 1983 as the base year.

(b) Determine the weighted aggregate volume index for 1984 and 1985 using the Paasche method with 1983 = 100.

(c) Determine Irving Fisher's ideal price index for 1985 using 1983 as base.

(d) Determine Drobish's volume index for 1985 using 1983 as the base year.

(e) With 1983 as the base year, develop the unweighted average of relatives price index for 1984.

Exercise 3 The North branch of the Pine Timber outlet has recorded sales for four of its outlets as follows:

Outlet	Sales (× R10 000)		
	1982	1983	1984
Pinetree	84.7	86.3	87.4
Elmwood	72.6	70.9	70.0
Oakmont	64.5	87.6	99.4

Compute the unweighted average of relatives price index for 1983 and 1984 with 1982 as the base year.

Exercise 4

1965		1970		1980	
Rent	No. of Tenants	Rent	No. of Tenants	Rent	No. of Tenants
0.50	1280	0.70	1 280	1.20	1 280
0.75	760	1.00	750	1.65	750
1.00	410	1.25	580	2.00	590
1.50	86	1.85	400	2.50	700
2.00	70	2.45	250	3.25	950
4.00	10	4.60	190	5.00	1 000

(a) Calculate Laspeyres and Paasche's price indices for 1970 and 1980 with 1965 = 100

(b) Calculate Irving Fisher and Drobish's price indices for 1970 and 1980.

Exercise 5 Mr Hiram has kept a record on the costs of certain items purchased weekly:

Item	Price per unit			Quantity purchased		
	1980	1981	1982	1980	1981	1982
Coffee	2.00	2.10	2.20	30	32	35
Cookies	0.60	0.65	0.70	6	7	7
Fruit	0.40	0.42	0.44	12	14	16
Sugar	2.10	2.40	1.80	20	22	25

(a) Construct a simple price index for coffee and sugar with 1980 = 100.

(b) Construct a simple quantity index for fruit with 1981 = 100.

(c) Calculate a composite price index for each year using the simple aggregate method with 1980 = 100.

(d) Calculate a composite volume index for each year using the simple aggregate method with 1980 = 100.

(e) Calculate a composite price index for each year using the unweighted average of relatives method and 1980 = 100.

(f) Calculate a composite quantity index for each year using the unweighted average of relatives method and 1980 = 100.

(g) Calculate a composite price index for each year using the weighted aggregate method and 1980 = 100.
  (i) Use Laspeyres weights.
  (ii) Use Paasche's weights.

INDEX NUMBERS

Exercise 6 Given the following product prices, compute the simple aggregate price index for 1985 using 1984 = 100.

Product	1984 (R)	1985 (R)
1	500	800
2	900	1 100
3	400	700
4	1 000	1 200

Exercise 7 Mr Rolling has been offered a new job with a salary of R23 500 per year in a city with a cost of living index of 162. If he presently earns R21 500 per year in a town with a cost of living index of 137, will he be better off financially in the new job?

Exercise 8 The Producer Commodity price index values for all commodities for the months January to December 1979, with 1967 = 100 were:
221 224 227 230 232 234 237 238 242 245 247 249
Shift the base of this index to May 1979.

Exercise 9 Monthly retail sales of cars (in millions of rands) for the first five months of 1980 were:
15.69 15.14 13.48 12.25 12.03
(*a*) Construct an index series showing changes in these sales during the months January 1980 to May 1980, with January = 100.
(*b*) Shift the base to March 1980.

Exercise 10 The Consumer Price Index values for the first eight months of 1980 with 1977 = 100 were:
233 236 240 243 248 249 252 255
(*a*) Shift the base of the index to June 1980.
(*b*) Determine the purchasing power of the rand for those months with 1977 = 100.

CHAPTER 14
Statistical Decision Theory

CONTENTS

14.1 INTRODUCTION 235
14.2 PROBLEM FORMULATION 236
14.3 CLASSIFICATION OF DECISION PROBLEMS 237
14.4 SUMMARIZING AND ANALYSIS OF A DECISION PROBLEM 237
14.5 DETERMINING THE BEST ACTION 239

14.1 INTRODUCTION

Chapter 8 explained the subject of hypothesis testing, the purpose being to help one reach a decision about a population by examining sample data from a population. This type of decision-making is known as classical statistical inference. Only two states of the parameter are being tested and economic consequences are not considered an explicit part of this decision procedure.

An alternative to classical statistical inference is statistical decision theory. Statistical data can be applied to achieve optimal or best decisions in an economic sense. This involves a logical and quantitative analysis of all factors and possibilities which can influence a decision problem and assists in arriving at an appropriate action to solve a problem. It is based upon the decision-maker's subjective or personal preferences and perceptions regarding evaluation of probabilities to be used in a decision framework. Such probabilities express the strength of one's belief with regard to the uncertainties that are involved when there is little or no direct information available.

The purpose of statistical decision theory is to develop and improve the decision-maker's understanding of applicable techniques to ensure that

- problems are formulated clearly,
- economic consequences are incorporated into the analysis,

❏ the likelihood of selecting an outcome or act from a set of possible acts, which will produce the best results in the context of the decision problem is increased.

Two assumptions underlying decision analysis are:

(*a*) that there is either a gain or a loss associated with each possible action available, and

(*b*) that the decision-maker selects that solution which maximizes the expected gain or minimizes an expected loss.

Decision Analysis generally includes the following steps:

❏ Problem formulation
❏ Summarizing and analysing of decision problem:
  − Construction of a decision (or payoff) table
  − Construction of a decision tree
  − Construction of an opportunity loss table
❏ Selecting the most favourable result in order to make a decision

14.2 PROBLEM FORMULATION

An individual can be said to have a problem only if he does not know what course of action is best and/or if he is in doubt about the solution.

To define and formulate a problem situation so that it is susceptible to research, the following conditions must be met:

❏ The decision-maker must have several possible options to evaluate prior to selecting one course of action.
❏ The decision-maker must be able to list all the possible conditions (states of nature or events) which will affect the outcome of each action.
❏ All the possible outcomes for each action pertaining to each state of nature must be measurable.
❏ The decision-maker must be able to determine how to select the best course of action or that alternative which results in the largest average profit.

However, when little or no information is available concerning the probabilities of occurrence of the various events, other criteria are available for making decisions.

14.3 CLASSIFICATION OF DECISION PROBLEMS

All decision problems can be classified according to one of three categories namely certainty, uncertainty or risk.

Decisions under conditions of certainty When the state of nature is known or has no influence on the outcome of actions, it is possible to calculate the outcome for each action, compare them, and select that one which gives the best result. The most common method for quickly identifying the best action, without having to consider every possible action, is by the method of linear programming. This method does not however fall within the ambit of statistics and hence is not covered in this text.

Decisions under conditions of uncertainty or risk There is a fine distinction between decision-making under conditions of uncertainty and decision-making under conditions of risk.

We are said to make a decision under uncertainty in situations where the outcome of each action will differ depending on the conditions which prevail and it cannot be predicted with certainty which conditions will prevail because the conditions (states of nature) are beyond the control of the decision-maker. Probabilities can therefore not be assigned to the possible outcomes at the time of decision-making.

Risk is a condition where the occurrence of the possible conditions can be assigned probabilities. It is known which conditions are possible and the probability of each event can be specified due to the availability of objective or historical data.

Should no objective data be available, the decision-maker may, by making use of Bayesian decision theory, assign probabilities to the states of nature to assist in making a decision (managerial judgement). Once probabilities are assigned, regardless of the manner in which they were obtained, the decision procedure that follows is exactly the same as for risk conditions.

14.4 SUMMARIZING AND ANALYSIS OF A DECISION PROBLEM

Constructing a decision or payoff table: (matrix format)

The following steps are to be observed:

- Step 1: List all actions which you might consider in order to solve the problem.
- Step 2: List all the states of nature which can occur for each alternative course of action.

- Step 3: The states of nature ($E_1, E_2 \ldots$) each indicate a separate column and the actions ($A_1, A_2, \ldots$) each indicate a separate row of the payoff table.
- Step 4: A payoff value is entered in each of the cells of the table. The payoffs or outcomes are those values or economic consequences available to the decision-maker should he take a specific action, assuming that each of the states of nature does occur.

Payoff table

Action	State of nature (E)				
	E_1	E_2	E_3	...	E_k
A_1					
A_2					
A_3					
.					
A_n					

If an action does not involve risk, the payoff will be the same no matter which state of nature occurs.

At least two alternative actions must be available so that a choice exists. If an action involves risk, the payoff table may also include the probability value for each event based either on historical information or managerial judgement. Due to the events in the payoff table being mutually exclusive as well as exhaustive, the sum of the probabilities for all the possible events must equal 1.0.

Construction of decision trees This is a graphical method which illustrates a decision problem by showing, in chronological order from left to right, every potential action, event and payoff. This method is particularly useful when a decision problem involves a sequence of many decisions.

- Step 1: Start on the left side of a page and work from left to right across the page. Use a small square to depict the base of the tree and from it draw lines to represent each action related to a decision problem. Branches representing actions (A), originate from small squares (□). The decision-maker begins by selecting an act from the set of acts.
- Step 2: Each branch leads towards the right and can be followed by another decision point (square) or by an event point. Create a branch for each state of nature (event) confronting the decision-maker. State of nature branches emanate from small circles.(O)

STATISTICAL DECISION THEORY

- Step 3: The outcome resulting from each act/state of nature combination is shown as the end position of the corresponding path from the base of the tree. All outcomes should be stated in terms of the same numerical measurement unit.
- Step 4: The decision-maker can proceed, working from the outcomes backwards towards the base of the tree, to find the decision that best satisfies the objectives.

Construction of an opportunity-loss table   Opportunity-Loss is simply the absolute value of the difference between the payoff actually realized for a selected act and the payoff which could have been obtained had the best or optimal act been selected.

From a payoff table we can construct an opportunity-loss table. An opportunity-loss table is one that shows the opportunity loss for every action-state of nature combination.

Calculating opportunity losses

Opportunity losses are calculated for each column of the opportunity-loss table as follows:
- Step 1: From the payoff table, find the highest payoff in each column.
- Step 2: Subtract each payoff from the highest payoff in the column in which the former payoff is located.

Note 1: The best action for each state of nature will have a zero entry.
Note 2: Entries in an opportunity-loss table will always be positive.
Note 3: Entries represent losses; therefore, the smaller the value of an entry, the better.

14.5 DETERMINING THE BEST ACTION

There are several methods which can be used to help to make a decision about the best action. We distinguish between methods for decision-making without probabilities and decision-making with probabilities.

Decision-making without probabilities   A prime consideration in choosing a method for determining the best decision when the probabilities about the events are unknown, is the decision-maker's general attitude towards possible losses and gains.

Maximin criterion   This procedure guarantees that the decision-maker can do no worse than to achieve the best of the poorest outcomes possible. This method will be the typical choice of a risk averter. To apply this criterion, determine the lowest (minimum) payoff associated with each action (smallest value of each row) and choose that action (row) with the largest minimum payoff. That means maximizing the minimum payoffs. This criterion applies only when the payoffs are quantities such as profit which is to be increased.

If the payoffs are quantities such as losses or costs, which are to be minimized, the minimax criterion is used instead.

Select the worst that can happen, that is the highest value for each row (action) in the payoff table. Choose that action (row) which minimizes this maximum payoff. This will be the action with the smallest maximum payoff.

Maximax criterion Using this criterion, the decision-maker is only concerned with the best that can happen with respect to each decision. It will be the typical choice of a risk seeker. To apply this criterion, find the highest value for each action (row) in the payoff table and select that action where maximum payoff is the highest. This means maximizing the maximum payoffs.

If the payoffs are quantities such as losses or costs to be minimized, use the minimum criterion instead. Determine and select the lowest possible cost (minimum) associated with each action, identify the lowest cost associated with these minima and then choose that action associated with the minimum of minima.

Minimax regret This criterion is applied by using an opportunity-loss table. A regret or opportunity loss equals the difference between a payoff and what it would have been if the best action for the given state of nature was selected. According to this criterion, the decision-maker finds the highest possible (maximum) regret value associated with each possible action (row), identifies the lowest (minimum) value amongst these maxima, and then chooses that action for which the maximum regret is smallest. In other words, minimize the maximum regret.

Decision-making with probabilities If the decision problem falls within the risk category, then the decision-maker can use probabilities to measure the likelihood of the occurrences of the various states of nature.

If the possible number of alternatives are limited, payoff tables and decision trees can be used to solve problems and discreet probability distributions will be applicable.

STATISTICAL DECISION THEORY

If the number of possibilities become too large, it is more realistic to analyse a decision in terms of continuous probability distributions such as the normal distribution.

USING PRIOR ANALYSIS TO DETERMINE THE BEST DECISION ACTION: DISCRETE PROBABILITY DISTRIBUTIONS

Prior probabilities are determined using information available from past experience, judgmental (subjective) information about the states of nature or a combination of available and subjective information. No investigation, sampling or other form of experiment concerning the data has been carried out in order to estimate these probabilities. The decision-maker arrives at conclusions based upon assumptions concerning the probabilities of the various outcomes as follows:

- Step 1: Specify the possible decision actions
- Step 2: Specify the possible states of nature
- Step 3: Set up a payoff table
- Step 4: Assume probabilities for each event
- Step 5: Compute the expected profit for each state of nature
- Step 6: Compare the expected profits and make a decision by choosing the highest expected payoff (expected monetary value criterion — EMV)

The expected monetary value (EMV) criterion This criterion is applied to select an action which has the highest expected payoff or monetary value (EMV) and is referred to as Bayes's Decision Rule.

Calculation of the EMV
- Step 1: A probability, which represents the likelihood of the occurrence of the states of nature, must be included in the payoff table.

Action	State of nature (E)				
	E_1	E_2	E_3	...	E_k
	P_1	P_2	P_3	...	P_k
A_1					
A_2					
A_3					
.					
A_n					

❑ Step 2: Multiply the payoff in each cell of the payoff table by the probability of the state of nature for that column.
❑ Step 3: The resulting expected payoffs are then summarized for each action to determine the EMV of the action.
❑ Step 4: Choose the maximum EMV.

To apply the EMV criterion using a decision tree, a process known as Backward Induction is used. This method implies that it is impossible to evaluate an immediate action without first considering all later decisions that will result from this choice. Evaluations must be made in reverse of their natural chronological order.

❑ Step 1: Calculate the monetary payoffs which are located at the right-hand side of the decision tree.
❑ Step 2: State the respective probabilities for each state of nature branch.
❑ Step 3: Determine the expected value associated with each of the decision alternatives by starting on the right-hand side and multiplying each monetary payoff with the probability of the state of nature for that branch-end.
❑ Step 4: Compare the expected values for the different decision actions and choose the best one. A double bar is entered for all other choices, indicating that they have been eliminated from further consideration. Branch elimination takes place only at action forks and never at states of nature forks.
❑ Step 5: Move to the left until the highest expected payoff is identified.

Minimum expected opportunity losses (EOL) The minimum expected opportunity loss measures the expected cost of uncertainty due to our uncertain knowledge about which state of nature will prevail.

The purpose is to select that decision action which yields the minimum expected opportunity loss (EOL). This will always lead to the same decision as the EMV criterion. The EOL for each action is calculated in the same way as the EMV.

The expected value of perfect information (EVPI) Before actually selecting a decision action, another problem faced by a decision-maker is to decide whether to stop the analysis after using only prior (already available) information or to postpone the final decision until additional sample information about the states of nature can be obtained in order to reduce uncertainty. To help in this decision, the decision-maker should be concerned about the cost of additional information and its potential value.

A measure to evaluate this cost is known as the expected value of perfect information (EVPI), which indicates the maximum expected gain in profit if additional information can provide certainty about the states of nature. Thus, the maximum amount of money that should be spent to know precisely the state of nature that will be in effect, should not exeed this maximum gain. With perfect information available, the decision-maker will know which state of nature will occur, eliminating uncertainty, and make a decision that is best for that state of nature.

To calculate the EVPI the opportunity loss table can be used and the expected opportunity loss (EOL) calculation. The EOL of an action is the difference between the expected profit when perfect information is available and the expected profit under uncertain conditions. The minimum EOL is thus the expected value of perfect information.

If the EVPI is high, then it will be wise to engage in sampling additional information. If additional information is not available at a cost equal to or less than the EVPI, the decision-maker cannot improve average profit by obtaining additional information.

USING POSTERIOR ANALYSIS TO DETERMINE THE BEST DECISION ACTION: DISCRETE PROBABILITY DISTRIBUTIONS

Posterior analysis is a form of decision-making that starts out with prior probabilities, proceeds to gather additional sample information about states of nature probabilities and then uses this new evidence to transform the set of prior probabilities, by means of Bayes's Theorem into a revised set of posterior probabilities.

Steps to follow in decision-making with sample information:

- Step 1: Specify the possible decision actions.
- Step 2: Specify the possible states of nature.
- Step 3: Set up the payoff table.
- Step 4: Assume prior probabilities for each event
- Step 5: Design and select a sample.
- Step 6: Determine conditional probabilities based upon additional information for the states of nature.
- Step 7: Compute joint and posterior probabilities for each state of nature. To calculate the posterior probability in the last column of the payoff table, the joint probability of each state of nature is divided by the sum of the joint probability column.
- Step 8: Use the posterior probabilities to compute the posterior expected profit for each decision action.

- Step 9: Compare the posterior expected profits and make a decision by determining the highest expected payoff.
- Step 10: Calculate the expected payoff of sampling.
- Step 11: Calculate the expected value of sample information.
- Step 12: Calculate the expected net gain for sampling.

The basic purpose of attempting to incorporate more evidence through sampling, is to reduce the expected cost of uncertainty. Suppose a decision-maker concludes that it will be worthwhile to defer a decision and first collect additional information about the states of nature, by making use of sampling (posterior analysis), before determining the best decision action. The expected monetary value decisions are formulated in the same way as before except that posterior probabilities are used instead of prior probabilities. Sampling has economic value only if the sample result causes a change in the best decision, as compared with the choice based on the prior probability distribution. If a posterior expected value is better than the prior expected value, but the decision is not changed, the sample adds no value because the higher expected value would have been experienced anyway but without knowledge of this fact.

Posterior analysis is based upon the application of Bayes theorem to revise the prior probabilities by making use of additional information obtained through sample results.

To apply Bayes Theorem the prior probability of the uncertain state of nature must be designated and the conditional probability of the sample result must be known.

State of nature	Prior probability P(A)	Conditional probability (P(B|A)	Joint probability (P(A).(P(B|A)	Posterior probability P(A|B)
E_1				
E_2				
E_3				
.				
.				
Total	1.00			1.00

Example 1   An entrepreneur planned to print souvenir T-shirts to sell during a big sporting event. A decision must be made whether to produce 10 000, 15 000 or 20 000 T-shirts. The demand for T-shirts will depend on attendance which may be high, medium or low with respective probabilities of 0.4, 0.35 and 0.25. The entrepreneur constructed the

STATISTICAL DECISION THEORY

following payoff table which shows the expected different profits in R'000 for attendance levels together with each quantity of T-shirts he can produce.

Payoff table

Actions	Attendance levels (states of nature)		
	High	Medium	Low
A_1 10 000	12	10	9.6
A_2 15 000	20	18	6
A_3 20 000	30	16	4

Decision tree

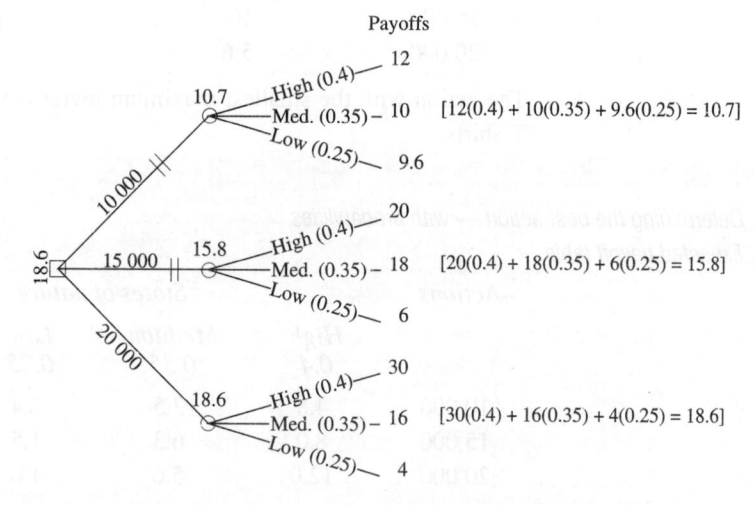

Payoffs

- 10 000 → 10.7
  - High (0.4) — 12
  - Med. (0.35) — 10 $[12(0.4) + 10(0.35) + 9.6(0.25) = 10.7]$
  - Low (0.25) — 9.6
- 15 000 → 15.8
  - High (0.4) — 20
  - Med. (0.35) — 18 $[20(0.4) + 18(0.35) + 6(0.25) = 15.8]$
  - Low (0.25) — 6
- 20 000 → 18.6
  - High (0.4) — 30
  - Med. (0.35) — 16 $[30(0.4) + 16(0.35) + 4(0.25) = 18.6]$
  - Low (0.25) — 4

18.6

Opportunity-loss table

Actions	States of Nature		
	High	Medium	Low
10 000	(30 – 12) = 18	(18 – 10) = 8	(9.6 – 9.6) = 0
15 000	(30 – 20) = 10	(18 – 18) = 0	(9.6 – 6) = 3.6
20 000	(30 – 30) = 0	(18 – 16) = 2	(9.6 – 4) = 5.6

Determining the best action — without probabilities
Maximin criterion

Action	Lowest payoff per action
10 000	9.6
15 000	6
20 000	4

The action with the largest minimum payoff is to produce 10 000 T-shirts.

Maximax criterion

Action	Largest payoff per action
10 000	12
15 000	20
20 000	30

The action with the highest maximum payoff is to produce 20 000 T-shirts.

Minimax regret

Action	Maximum regret from opportunity-loss table
10 000	18
15 000	10
20 000	5.6

The action with the smallest maximum regret is to produce 20 000 T-shirts.

Determining the best action — with probabilities
Expected payoff table

Actions	States of nature			
	High 0.4	Medium 0.35	Low 0.25	Total
10 000	4.8	3.5	2.4	10.7
15 000	8.0	6.3	1.5	15.8
20 000	12.0	5.6	1.0	18.6

The maximum Expected Monetary Value (EMV) is 18.6 and the decision will be to produce 20 000 T-shirts.

Expected opportunity-loss table (EOL)

Actions	States of nature			
	High 0.4	Medium 0.35	Low 0.25	Total
10 000	(18 × 0.4) = 7.2	(8 × 0.35) = 2.8	(0 × 0.25) = 0.0	10.0
15 000	(10 × 0.4) = 4.0	(0 × 0.35) = 0.0	(3.6 × 0.25) = 0.9	4.9
20 000	(0 × 0.4) = 0.0	(2 × 0.35) = 0.7	(5.6 × 0.25) = 1.4	2.1

Since we are dealing with losses we would select the action that minimizes the expected opportunity loss, namely to produce 20 000 T-shirts.

STATISTICAL DECISION THEORY

The expected value of perfect information (EVPI) It can be mathematically proved that EVPI = EOL, therefore the maximum amount the entrepreneur would be willing to pay for additional sample information is R2.1 thousand because that is the maximum expected gain in profits that could result if more accurate information is available.

DETERMINING THE BEST ACTION IF SAMPLE INFORMATION IS KNOWN

A research firm who charges R1 000 to conduct a survey, provides the following data to estimate attendance levels.

In the past when such studies predicted high attendance, it was high 82 %, medium 9 % and low 3 % of the time. When medium attendance was predicted, it was high 15 %, medium 74 % and low 12 % of the time. When low attendance was predicted it was high 3 %, medium 17 % and low 85 % of the time.

Calculation of joint probabilities
(prior probabilities × conditional probabilities)

	Prior	High|E_i	Joint	Med|E_2	Joint	Low|E_3	Joint
E_1	0.4	0.82	0.328	0.15	0.06	0.03	0.012
E_2	0.35	0.09	0.0315	0.74	0.259	0.17	0.0595
E_3	0.25	0.03	0.0075	0.12	0.03	0.85	0.2125
			0.3670		0.349		0.2840

Calculation of posterior probabilities

	E_i|high	E_2|med	E_3|low
E_1	0.328/0.367 = 0.8937	0.06/0.349 = 0.1719	0.012/0.284 = 0.0423
E_2	0.0315/0.367 = 0.0858	0.259/0.349 = 0.7421	0.0595/0.284 = 0.2095
E_3	0.0075/.367 = 0.0204	0.03 /0.349 = 0.0860	0.2125/0.284 = 0.7482
	1.0000	1.0000	1.0000

Calculation of expected payoffs for each action:
(Payoff × posterior probability)

For action 1 (10 000 T-shirts)

High demand	Medium demand	Low demand
12(0.8937) = 10.7244	12(0.1719) = 2.0628	12(0.0423) = 0.5076
10(0.0858) = 0.858	10(0.7421) = 7.421	10(0.2095) = 2.095
9.6(.0204) = 0.1958	9.6(.0860) = 0.8256	9.6(.7482) = 7.1827
11.7782	10.3094	9.7853

For action 2 (15 000 T-shirts)

High demand	Medium demand	Low demand
20(0.8937) = 17.874	20(0.1719) = 3.438	20(0.0423) = 0.846
18(0.0858) = 1.5444	18(0.7421) = 13.3578	18(0.2095) = 3.771
6 (0.0204) = 0.1224	6(0.0860) = 0.516	6(0.7482) = 4.4892
19.5408	17.3118	9.1062

For action 3 (20 000 T-shirts)

High demand	Medium demand	Low demand
30(0.8937) = 26.811	30(0.1719) = 5.157	30(0.0423) = 1.269
16(0.0858) = 1.3728	16(0.7421) = 11.8736	16(0.2095) = 3.352
4(0.0204) = 0.0816	4(0.0860) = 0.344	4(0.7482) = 2.9928
28.2654	17.3746	7.6138

Summary of posterior expected payoffs

	E_1	E_2	E_3
Joint probability	0.3670	0.3490	0.2840
A_1	11.7782	10.3094	9.7853
A_2	19.5408	17.3118	9.1062
A_3	28.2654	17.3746	7.6138

The maximum expected monetary value with sample information is:

EMS = Σ(maximum expected payoff per column × joint
     probability of that state of nature)
    = 28.2654(0.3670) + 17.3746(0.3490) + 9.7853(0.2840)
    = 19.2162

The expected monetary value (EMV) without sample information was 18.6. To calculate the expected value of sample information:

EVSI = EMS – EMV
     = 19.2162 – 18.6
     = 0.6162 or R616.20

The expected gain that could result from sampling then is R616.20.

To decide whether the entrepreneur should hire the research firm the expected net gain through sampling should be calculated.

ENGS = EVSI – Cost of sampling
     = R616.20 – R1 000
     = –R338.80

Since the ENGS is negative, the entrepreneur should not pay for sample information.

Example 2 Kornles must determine whether or not to market a new breakfast food. A decision must also be made whether to conduct a consumer test market programme at a cost of R50 000. If the new product is successful, profits will increase by R2 000 000, whereas a failure will cause a loss of R1 000 000 to the company. Not marketing the product would not change profits. Without testing, the product has been judged to have a 60 % chance of success. If the market is being tested, the assumed probability of a favourable test result is 52 %. Given a favourable test result, the chance of success is judged to be 80 %. If the results are unfavourable, the product's success probability is judged to be only 10 %. Construct a decision tree diagram to determine which action will provide the greatest expected payoff, including the choice of whether or not to test.

ACTION   STATES OF NATURE

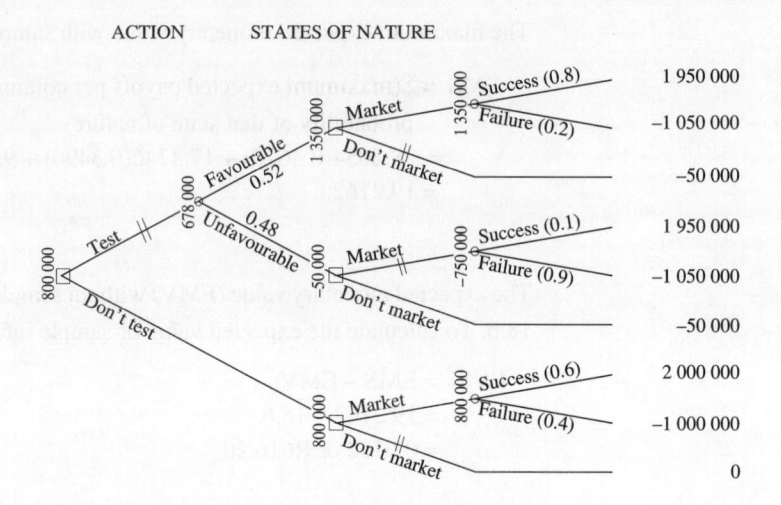

[Decision tree diagram:
- Test (678 000) / Don't test (800 000)
- Test branch:
  - Favourable 0.52 (1 350 000)
    - Market (1 350 000): Success (0.8) → 1 950 000; Failure (0.2) → −1 050 000
    - Don't market → −50 000
  - Unfavourable 0.48 (−50 000)
    - Market (−750 000): Success (0.1) → 1 950 000; Failure (0.9) → −1 050 000
    - Don't market → −50 000
- Don't test (800 000):
  - Market (800 000): Success (0.6) → 2 000 000; Failure (0.4) → −1 000 000
  - Don't market → 0]

The best action is to market without consumer testing. The expected payoff then is R800 000.

EXERCISES

Exercise 1 Square Eye TV film producer wants to market a new film. The rights can be sold to a distributor for R125 000 now, or the production can be offered to a TV network for review with the following results according to the producer: A 60 % chance of rejection with a resulting loss of R30 000; a 40 % chance of getting the contract which means a R300 000 profit. A consultant is willing to offer advise on the network's likely reaction for a R1 000 fee. The consultant's survey indicates the following results:

Advice given (A)	Rejection (E_1) $P(A_i\|E_1)$	Contract offer (E_2) $P(A_i\|E_2)$
Network will reject	0.8	0.3
Contract offer	0.2	0.7

Do a decision analysis for the film producer.

Exercise 2 The demand for a new route offered by an airline company is somewhat uncertain. You are required to decide which type of aeroplane to use on this route: the 747 with a 350 passenger capacity or the 707 with a 250 passenger capacity. The probability of high annual ticket sales is 0.10 with a resulting profit of R1 000 000 with the 747 and R400 000 with the 707. The probability of medium ticket sales is 0.40 with a resulting profit of R500 000 with the 747 and R300 000 with

STATISTICAL DECISION THEORY

the 707. The probability of low ticket sales is 0.30 with a resulting profit of R300 000 with the 747 and R200 000 with the 707. The chance of poor ticket sales is 0.20 with a resulting loss of R300 000 with the 747 and a profit of R50 000 with the 707.

A consultant was hired to determine the expected ticket sales for the new route. The sample results indicated medium sales and in the past when such studies indicated medium sales, sales actually have been medium 45 % of the time, high 5 % of the time, low 35 % of the time and poor 15 % of the time.

Exercise 3 A group of investors are interested in investing R500 000 in one of three investments, which are affected by the price of gold, for a period of one year. The alternatives are gold, platinum and financial rand. The return (in R'0000) for each investment versus the fluctuation of gold prices during the year is as follows:

Fluctuation of gold prices	Investment alternatives		
	Gold	Platinum	Fin Rand
Up	16	−2	−10
Unchanged	−6	−1	20
Down	4	5	6

Based on past experience, the following probabilities are assigned for the price of gold during the year:
Up	0.55
Unchanged	0.30
Down	0.15

Before making the final decision, the investors are considering an economic forecast for the forthcoming year. Based on past forecasts, when the price of gold has gone up, the probability of a good economic forecast was 0.80; when the price of gold has stayed the same, the probability of a good forecast was 0.50; when the price of gold has gone down the probability of a good economic forecast was 0.24. If the forecast indicates a bad economy, revise the prior probabilities in the light of the new information.

Exercise 4 You must decide whether or not to buy a manufacturing operation of one of your competitors. The choices are to buy the complete operation, only the property, or not to buy anything. The results are indicated in the following payoff table in rands:

Demand	Decision		
	Complete company	Property only	Nothing
High	200 000	100 000	20 000
Medium	120 000	75 000	60 000
Low	50 000	40 000	140 000

Assume that the probabilities of high, medium and low demand are 0.4, 0.5 and 0.1 respectively.

Economic forecasters anticipate high demand. In the past if high demand was forecasted, the demand was high 50 % of the time, medium 30 % of the time and low 20 % of the time.

CHAPTER 15
Interest Calculations

CONTENTS

15.1 INTRODUCTION 253
15.2 SIMPLE INTEREST 253
15.3 COMPOUND INTEREST 255
15.4 NOMINAL AND EFFECTIVE RATES OF INTEREST 257
15.5 ANNUITIES CERTAIN 258

15.1 INTRODUCTION

Interest is money paid for the use of borrowed money or money earned when capital is invested.

The sum on which the interest is calculated is called the principal or present value.

The rate is the percentage of the principal that is to be paid for each unit of time and is expressed as a percentage per year.

The time is the period for which the money is borrowed or invested and is expressed in years or a fraction of a year.

The amount or future value at any time is the sum of the principal and the interest that has accrued up to that time.

15.2 SIMPLE INTEREST

Where, for the whole term of the contract, only the original principal earns interest, it is called simple interest and is defined as the principal × rate × time.

$$I = Prt$$
$$S = P(1 + rt)$$

I = simple interest
P = present value

r = rate per year
t = time in years
S = amount or future value

Exact simple interest is calculated on a basis of 365 days per year (366 in a leap year).
Ordinary simple interest is calculated on a basis of 360 days per year or 30 days per month.

Example 1   Determine the exact simple interest on R2 000 for 50 days at a rate of 5 %.

$$I = Prt$$
$$= 2\,000\,(0.05)\,\frac{50}{365}$$
$$= R13.70$$

Example 2   Determine the amount in the previous example.

$$S = P + I$$
$$= 2\,000 + 13.70$$
$$= R2\,013.70$$

Example 3   Determine the present value at 15 % simple rate of interest of R12 500 due in 1 year and 9 months.

$$S = P(1 + rt)$$
$$12\,500 = P[1 + 0.15\,(1.75)]$$
$$P = R9\,900.99$$

Example 4   B borrows R500 from A and at the end of 1 year pays A a sum of R525. What is the simple rate of interest earned?

$$I = Prt$$
$$25 = 500r(1)$$
$$r = 0.05 = 5\%$$

Example 5   How long will it take R5 000 to earn R50 interest at 10 %?

$$I = Prt$$
$$50 = 5\,000\,(0.10)t$$
$$t = 0.1$$

0.1 of 1 year is 1 month and 6 days.

INTEREST CALCULATIONS

15.3 COMPOUND INTEREST

When interest is not paid out, but continuously added to the principal, the principal is continuously increasing and the interest is said to be compounded.

$$S = P(1 + i)^n$$

i = the rate of interest per period
n = the number of interest periods per transaction

The answer can be determined by using either the formula or the compound interest tables.

Example 1   If R1 000 is invested for 8 years at 6 % compounded quarterly, what sum of money will be in the bank at the expiry date?

$i = 1.5$ % per quarter
$n = 32$ quarters
$S = P(1 + i)^n$
$= 1\,000\,(1 + 1.5\,\%)^{32}$
$= 1\,000\,(1.6103)$ [value obtained from compound interest table $(1 + i)^n$]
$= R1\,610.30$

Example 2   A young man inherited R200 000. He wants to invest a portion of his inheritance. His goal is to accumulate R300 000 in 15 years. What portion of the money should be invested if the money will earn 8 % per year compounded semiannually. How much interest will be earned over the period?

$i = 4$ % per half year
$n = 30$ half years
$S = P(1 + i)^n$
$300\,000 = P(1 + 4\,\%)^{30}$
$= P(3.2433)$ [value obtained from compound interest table $(1 + i)^n$]
$P = R92\,498.38$

$I = S - P$
$I = 207\,501.62$

Example 3   If R100 is invested at 5 % compounded quarterly for two years and eleven months, how much will be in the account?

$i = 1\tfrac{1}{4}$ % per quarter

$n = 11$ quarters and 2 months

$S = P(1 + i)^n$

$= 100(1 + 1\frac{1}{4})^{11}$ + simple interest for 2 months

$= 100(1.1464)$ + simple interest for 2 months

$= 114.64 + \left[114.64 \times 0.05 \times \dfrac{2}{12} \right]$

$= R115.60$

Example 4   What sum of money was invested two years and ten months ago if R1 000 has been accumulated at 14 % interest compounded quarterly? When working from a known future value (S) back to an unknown present value (S) and the period given is an odd number, then the value of n must be equal to the next whole number. Simple interest for the difference between the whole number (n) and the original (odd) period is calculated and added on to the answer.

$i = 3\frac{1}{2}$ % per quarter

$n = 11$ quarters and 1 month

$S = P(1 + i)^n$

$1\,000 = P(1 + 3\frac{1}{2})^{12}$ plus simple interest for 2 months

$= P(1.5110)$ plus simple interest for 2 months

$P = 661.81$ plus simple interest for 2 months

$I = Prt$

$= 661.81 \times 0.14 \times \frac{2}{12}$

$= 15.44$

$P = R661.81 + 15.44$

$= R677.25$

Example 5   Determine the interest rate on a bond which would increase its value from R360 000 to R500 000 in 5 years if the interest is compounded monthly.

$S = P(1 + i)^n$

$500\,000 = 360\,000(1 + i)^{60}$

$1.3888 = (1 + i)^{60}$

The row in the table where $n = 60$ shows 1.3888 between:

1.3488	1.3888	1.5656
½ %	?	¾ %

Determine i by making use of interpolation.

INTEREST CALCULATIONS

$$\frac{1.3888 - 1.3488}{1.5656 - 1.3488} = \frac{0.04}{0.2168}$$

$$= 0.1845$$

$$i = \tfrac{1}{2}\% + 0.1845\,(\tfrac{3}{4}\% - \tfrac{1}{2}\%)$$

$$= 0.55$$

$$r = 0.55 \times 12$$

$$= 6.6\ \%\text{ per year}$$

Example 6   How long will it take for R20 to amount to R30 at 5 % compounded quarterly?

$$i = 1\tfrac{1}{4}\ \%\ \text{per quarter}$$

$$S = P(1 + i)^n$$

$$30 = 20(1 + 1\tfrac{1}{4}\ \%)^n$$

$$1.5 = (1 + 1\tfrac{1}{4}\ \%)^n$$

The $1\tfrac{1}{4}$ % interest column shows:

1.4881	1.5	1.5067
32	?	33

$$\frac{1.5 - 1.4881}{1.5067 - 1.4881} = \frac{0.0119}{0.0186}$$

$$= 0.6398$$

$$n = 32 + 0.6398(33 - 32)$$

$$= 32.64\ \text{quarters}$$

$$= 8\ \text{years, 1 month and 28 days}$$

15.4 NOMINAL AND EFFECTIVE RATES OF INTEREST

When interest is compounded more often than once a year, the given annual rate is called the nominal rate (i).

The interest actually earned per year, calculated as a rate, is known as the effective rate of interest (r).

$$\boxed{r = (1 + i)^m - 1}$$

where m is the frequency of conversion.

Example   What is the effective rate of interest equivalent to 10 % compounded monthly?

$$r = \left(1 + \frac{10}{12}\%\right)^{12} - 1$$
$$= (1.008333)^{12} - 1$$
$$= 0.1047$$
$$= 10.47\%$$

15.5 ANNUITIES CERTAIN

An annuity is a sequence of payments made at equal intervals of time and usually equal in amount.

An ordinary annuity is an annuity in which the payments begin and end on fixed dates.

A deferred annuity is one in which the term depends upon some event of which the occurrence cannot be fixed, such as life insurance, pensions, etc.

The time between successive payments (R) is called the payment interval and the time between the first payment and the last payment is called the term of the annuity.

The payment interval and the interest period always coincide.

Ordinary annuities certain An ordinary annuity certain is one in which the payments are made at the end of the payment intervals.

The amount is the value of the annuity on the day of the last payment.

$$S = R \cdot s_{\overline{n}|i}$$
$$= R \cdot \frac{(1+i)^n - 1}{i}$$

The present value of an annuity is an amount of money today which is equivalent to a series of equal payments in the future.

$$P = R \cdot a_{\overline{n}|i}$$
$$= R \cdot \frac{1 - (1+i)^{-n}}{i}$$

Example 1 Determine the amount and present value of an annuity certain of R150 per month for three years if money is worth 12 % compounded monthly. $n = 36$; $i = 1\%$

$$S = R \cdot s_{\overline{n}|i}$$

INTEREST CALCULATIONS

$$= 150(43.0768)$$
$$= R6\,461.52$$
$$P = R \cdot a_{\overline{n}|\,i}$$
$$= 150(30.1075)$$
$$= R4\,516.12$$

Example 2 A donor wants to provide a R3 000 scholarship every year for four years with the first to be awarded one year from now. If the school can get an 8 % return on investment, how much money should the donor give now?

$$= 3\,000\ a_{\overline{4}|\,8}$$
$$= 3\,000\,(3.3121)$$
$$= R9\,936.30$$

Example 3 Arthur Smith wants to have R6 000 in the bank in five years' time. He plans to deposit the correct amount to achieve this, at the end of each month. What should be the value of monthly payments if interest is 15 % compounded monthly?

$$S = R \cdot s_{\overline{n}|\,i}$$
$$6\,000 = R \cdot s_{\overline{60}|\,1\frac{1}{4}}$$
$$= R \cdot (88.5745)$$
$$R = 67.74$$

Example 4 A family buys a refrigerator that sells for R350 cash. They pay R50 deposit and the balance in 24 equal monthly payments. If the seller charges 18 % converted monthly, how much will the monthly payments be?

$$P = R \cdot a_{\overline{n}|\,i}$$
$$300 = R \cdot a_{\overline{24}|\,1\frac{1}{2}}$$
$$= R(20.0304)$$
$$R = 14.98$$

Annuities due An annuity due is an annuity whose periodic payment falls at the beginning of the payment interval, the first payment being due at once.

$$S = R(s_{\overline{n+1}|\,i} - 1)$$
$$= R \cdot \frac{[(1+i)^n - 1][1+i]}{i}$$
$$P = R(a_{\overline{n-1}|\,i} + 1)$$
$$= R \cdot \frac{[1 - (1+i)^{-n}][1+i]}{i}$$

Example 1 An investment of R200 is made at the beginning of each year for 10 years. If interest is 12 % how much will the investment be worth at the end of 10 years?

$$S = R(s_{\overline{n+1}|\,i} - 1)$$
$$= 200(s_{\overline{11}|\,12} - 1)$$
$$= 200(20.6545 - 1)$$
$$= 3930.90$$

Example 2 The premium on a life insurance policy is R60 per quarter, payable in advance. Determine the cash equivalent of a year's premiums if the insurance company charges 10 % converted quarterly for the privilege of paying this way instead of all at once for the year?

$$P = R(a_{\overline{n-1}|\,i} + 1)$$
$$= 60(a_{\overline{4-1}|\,2\frac{1}{2}} + 1)$$
$$= 60(2.8560 + 1)$$
$$= 231.36$$

Example 3 The beneficiary of a life insurance policy may take R10 000 in cash or 10 equal payments, the first to be made immediately. What is the annual payment if money is worth 12 %?

$$P = R(a_{\overline{n-1}|\,i} + 1)$$
$$10\,000 = R(a_{\overline{10-1}|\,12} + 1)$$
$$= R(5.3282 + 1)$$
$$R = 1580.23$$

Example 4 A student wants to save a R1 000 for a trip after graduation 4 years from now. How much must she save at the beginning of each month starting now if she gets 18 % compounded monthly on her savings?

$$S = R(s_{\overline{n+1}|\,i} - 1)$$
$$1\,000 = R(s_{\overline{48+1}|\,1\frac{1}{2}} - 1)$$
$$= R(71.6086 - 1)$$
$$R = 14.16$$

Deferred annuities A deferred annuity is one in which the first payment is made not at the beginning or end of the first period, but at some later date. When the first payment is made at the end of 10 periods, the annuity is said to be deferred for 9 periods. Similarly, an annuity that is deferred for 12 periods will have the first payment made at the end of 13 periods.

INTEREST CALCULATIONS

$$P \text{ (def)} = R(a_{\overline{m+n}|\,i} - a_{\overline{m}|\,i})$$

Where:
n = number of payments
m = number of deferred periods

Example 1 Determine the present value of a deferred annuity of R500 a year for 10 years, that is deferred 5 years. Money is worth 6 %.

$P \text{ (def)} = R\,(a_{\overline{m+n}|\,i} - a_{\overline{m}|\,i})$
$= 500\,(a_{\overline{5+10}|\,6} - a_{\overline{5}|\,6})$
$= 500\,(9.7122 - 4.2124)$
$= 2749.90$

Example 2 Determine the present value of an annuity of R50 every 3 months for 5 years if the first payment is made in three years. Money is worth 5 % converted quarterly.

$n = 20$
$R = 50$
$i = 1\frac{1}{4}\,\%$ per quarter
$m = 11$

$P(\text{def}) = 50(a_{\overline{31}|\,1\frac{1}{4}\%} - a_{\overline{11}|\,1\frac{1}{4}\%})$
$= 50(25.5693 - 10.2178)$
$= 767.58$

Example 3 A woman inherits R20 000. Instead of taking the cash, she invests the money at 3 %, converted semi-annually, on the understanding that she will receive 20 equal semi-annual payments of which the first payment will be made in 5 years. Determine the size of the payments.

$P \text{ (def)} = R(a_{\overline{m+n}|\,1\frac{1}{2}\%} - a_{\overline{m}|\,1\frac{1}{2}\%})$

$20\,000 = R(a_{\overline{20+9}|\,1\frac{1}{2}\%} - a_{\overline{9}|\,1\frac{1}{2}\%})$

$= R(23.3761 - 8.3605)$
$R = 1\,331.95$

EXERCISES

Exercise 1   Determine the simple interest on R750 for two months at 7 %.

Exercise 2   A woman borrows R30 000 to buy a home. The interest rate is 12 % and the monthly payment is R308.59. How much of the first payment goes to interest and how much to principal?

Exercise 3   A worker borrowed R150 for two months and paid R9.00 interest. What was the annual rate?

Exercise 4   If a person saves R3 000 at 10 %, how long will it take him to get R75 interest?

Exercise 5   How long will it take R625 to earn R25 interest at 4.8 %?

Exercise 6   A mechanic borrowed R125 from a loan company and at the end of one month paid off the loan with R128.75. What annual rate of interest was paid?

Exercise 7   What is the amount if R3.6 million is borrowed for one month at 2.875 %?

Exercise 8   A man borrows R95. Six months later he repays R100, which covers the loan, principal and interest. What interest rate did he pay?

Exercise 9   A waitress who was temporarily pressed for funds pawned her watch and diamond ring for R55. At the end of 1 month she redeemed them by paying R59.40. What was the annual rate of interest?

Exercise 10   How long will it take any sum of money to double itself at 5 % simple interest?

Exercise 11   What sum invested today at 15 % will amount to R1 000 in 8 months?

Exercise 12   A 10-month note for R3 000 with interest at 6 % was written today. Determine its value 4 months from today, if money is then worth 5 %.

Exercise 13   Determine the compound amount and the compound interest if R1 000 is invested for 10 years at 7 %?

Exercise 14   Determine the compound amount and the compound interest if R24 500 is borrowed for 3 years at 9 % converted monthly.

Exercise 15   The day a boy is born, his father invests R200 at 5 % compounded semi-annually. Determine the value of the fund on the boy's eighteenth birthday.

INTEREST CALCULATIONS

Exercise 16   The sales of a business have been increasing at the rate of 3 % per year. If the sales in 1981 were R250 000, what are the estimated sales to the nearest thousand rands for 1986?

Exercise 17   On 1 June 1977 a woman incurs a debt of R3 000 that is to be repaid on demand of the lender with interest at 9 % converted semi-annually. If the lender demands payment on 1 December 1985, how much must the borrower repay?

Exercise 18   As part of her retirement programme, a woman puts R12 000 in a savings account paying 8 % converted quarterly. The investment is made on the day she becomes 57. What will the maturity value of this account be if it matures when she becomes 62?

Exercise 19   What amount of money will be required to repay a loan of R1 835.50 on 1 July 1993, if the loan was made on 1 October 1989 at an interest rate of 8 % compounded quarterly?

Exercise 20   An investment of R4 000 is made for 12 years. During the first five years the interest rate is 9 % converted semi-annually. The rate drops to 8 % converted semi-annually for the remainder of the time. What is the final amount?

Exercise 21   On 1 June 1977 Charles Moser borrows R3 000 at 8 % converted semi-annually. How much does he have to repay on 28 September 1984?

Exercise 22   If R6 000 is borrowed for 5 years and 4 months at 9 % converted semi-annually, what amount would be required to repay the debt?

Exercise 23   What is the amount of R40 000 for 6 years and 3 months at 10 % converted semi-annually?

Exercise 24   Determine the compound amount of R1 000 for 8 years and 5 months at 4 % compounded semi-annually?

Exercise 25   Determine the compound amount of R1 500 for 6 years and 10 months at 5 % compounded quarterly.

Exercise 26   Determine the present value of R5 000 due in 4 years if money is worth 8 % compounded semi-annually.

Exercise 27   What principal is needed to accumulate R3 000 in 8 years at 6 % converted semi-annually?

Exercise 28   What is the present value at 10 % compounded quarterly of R12 000 due in 18 months?

Exercise 29   A person owns a note for R2 500 due in 5 years. What should a buyer, wishing money to earn 6 % converted quarterly, pay for the note?

Exercise 30   Determine the present value of R3 500 due in 4 years if money is worth 8 % compounded yearly?

Exercise 31   A note with a maturity value of R1 000 is due in 3 years and 8 months. What is the present value at 6 % compounded semi-annually?

Exercise 32   On 9 August 1983 Miss Lufkin borrowed R4 000 from Mrs Feld. She gave Mrs Feld a note promising to repay the money in 5 years with interest at 11 %. On 9 February 1985 Mrs Feld sold the note to a buyer who charged a rate of 13 % compounded semi-annually. How much did Mrs Feld get for the note?

Exercise 33   How much must X invest today at 4 % compounded quarterly to have R15 000 in his account 10 years from today?

Exercise 34   If R500 amounts to R700 in 5 years with interest compounded quarterly, what is the rate of interest?

Exercise 35   For a sum of money to double itself in 15 years, what must be the rate of interest converted semi-annually?

Exercise 36   The net income of the General Electric Company increased at approximately a constant rate from R471.8 million in 1981 to R608.1 million in 1984. Determine the annual rate.

Exercise 37   Personal consumption expenses went from R430.4 billion in 1975 to R1 670.1 billion in 1980.
Determine the annual compounded rate of growth.

Exercise 38   If R750 accumulates to R1 200 in 6 years, determine the nominal rate converted monthly.

Exercise 39   How long will it take R200 to amount to R350 at 7 % compounded semi-annually?

Exercise 40   How long will it take an investment to increase at least 50 % in rand value if the interest rate is 5 % converted annually?

INTEREST CALCULATIONS

Exercise 41 A loan company charges 36 % a year compounded monthly on small loans. How long will it take to triple money at this rate? Give the answer to the nearest month.

Exercise 42 A child, 8 years old, is left R1 600. Under the conditions of the will, the money is to be invested until it amounts to R2 500. If the money is invested at 6 %, how old will the child be when the R2 500 is received?

Exercise 43 A man invests R1 000 at 6 % compounded semi-annually on June 1, 1983. No interest is allowed for part of the period. How much will be in the account on 1 January 1990?

Exercise 44 What is the effective rate of interest equivalent to 7 % converted:
(a) semi-annually;
(b) quarterly; and
(c) monthly?

Exercise 45 Which gives the better annual return on investment, 4 % compounded quarterly or 4 % converted semi-annually? Show the figures on which you base your answer.

Exercise 46 A principal of R6 500 earns 5 % effective for three years and then 6 % compounded semi-annually for 4 more years. What is the amount at the end of the seven years?

Exercise 47 Determine the amount of an annuity of R5 000 per year for 10 years at (a) 6 %; and (b) 7 %

Exercise 48 Every 3 months a person puts R100 in a savings account that pays 5 % compounded quarterly. If the first deposit is made on 1 June 1979, how much will be in the account just after the deposit is made on 1 December 1988?

Exercise 49 What is the amount of an annuity of R100 at the end of each month for 6 years if money is worth 6 % compounded monthly?

Exercise 50 To provide for his son's education, a man deposits R150 at the end of each year for 18 years. If the money draws 6 % interest, how much does the fund contain just after the eighteenth deposit is made? If no more deposits are made but the amount in the fund is allowed to accumulate at the same interest rate, how much will the fund contain in 3 more years?

Exercise 51 R200 at the end of each year for 6 years is equivalent to what single payment at the end of 6 years if the interest rate is 6 % effective?

Exercise 52 A family has been paying R80 a month on their home. The interest rate on the bond is 9 % compounded monthly. Because of sickness they missed the payments due on 1 May, 1 June, 1 July and 1 August. On 1 September they want to make a single payment that will reduce their debt to what it would have been had they made all their payments on time. What single payment on this date will be equivalent to the 5 payments from May to September inclusive?

Exercise 53 A person has an income of R250 every 3 months from preferred stock. This income is deposited in a savings account paying 6 % converted quarterly with interest dates on 31 March, 30 June, 30 September, and 31 December. If the first deposit is made on 30 June 1981 and the last is made on 31 December 1985, how much will be in the account just after the last deposit?

Exercise 54 A donor wants to provide a R3 000 scholarship every year for 4 years with the first to be awarded 1 year from now. If the college can get 9 % return on investment, how much money should the donor give now?

Exercise 55 Determine the present value of an ordinary annuity of R1 200 at the end of each 6 months for 5 years if money is worth:
(a) 5 %;
(b) 6 %;
(c) 7 %
all compounded semi-annually?

Exercise 56 A television set is bought for R50 cash and R18 a month for 12 months. What is the equivalent cash price if the rate is 30 % converted monthly?

Exercise 57 A refrigerator can be purchased for R57.47 down and R20 a month for 24 months. What is the equivalent cash price if the rate is 30 % converted monthly?

Exercise 58 A home was purchased for R6 000 down and R200 a month for 20 years. If the monthly payments are based on 9 % converted monthly, what was the cash price of the house?

INTEREST CALCULATIONS

Exercise 59   If money is worth 5 % compounded quarterly, what single payment today is equivalent to 15 quarterly payments of R100 each, the first due 3 months from today?

Exercise 60   If money is worth 5 % compounded semi-annually, how much must a person save every 6 months to accumulate R3 000 in 4 years?

Exercise 61   A family wants to buy a home costing R60 000. If they pay R10 000 cash, how much will their monthly payments be if they get a 20 year bond with a rate of 9 % converted monthly?

Exercise 62   A person wants to accumulate R3 000 in 6 years. Equal deposits are made at the end of each 6 months in a savings account paying 6 % converted semi-annually. What is the size of each deposit?

Exercise 63   A loan company charges 36 % converted monthly for small loans. What would be the monthly payment if a loan of R150 is to repaid in 12 payments?

Exercise 64   On 1 June 1984, a widow takes the R10 000 benefit from her husband's insurance policy and invests the money at 6 % converted monthly with the understanding that she will receive 120 equal monthly payments. If she is to get her first cheque on 1 July 1984, what will be the size of each monthly payment?

Exercise 65   Determine the quarterly payments necessary for an ordinary annuity to amount to R4 500 in 4 years at 6 % converted quarterly.

Exercise 66   An investment of R200 is made at the beginning of each year for 10 years. If interest is 6 % effective, how much will the investment be worth at the end of 10 years?

Exercise 67   A student wants to have R600 for a trip after graduation 4 years from now. How much must she invest at the beginning of each year starting now if she gets 5 % compounded annually on her savings?

Exercise 68   The premium on a life insurance policy is R60 a quarter, payable in advance. Determine the cash equivalent on a year's premiums if the insurance company charges 6 % converted quarterly for the privilege of paying this way instead of all at once for the year?

Exercise 69   On 6 July 1983 a man deposits R250 in a savings account paying 5 % converted semi-annually. Interest dates are 30 June and 31 December. Deposits made by the 10th of the month can earn interest for the entire month. The man continues to make deposits of R250 every 6 months up to 4 January 1989, when he makes the last deposit. How much will be in his account after interest is credited on 30 June 1989?

Exercise 70 The premium on a life insurance policy is R15 a month payable in advance. A policyholder may pay a year's premium in advance. Determine the annual premium if it is based on 6 % converted monthly?

Exercise 71 The annual premium on a life insurance policy is R185 payable in advance. What would be the monthly premium based on 6 % converted monthly?

Exercise 72 A store can be rented for R350 a month payable in advance. The landlord will accept a single cash payment at the beginning of a year for a year's lease if interest is computed at 6 % converted monthly. What would be the cost of a year's lease?

Exercise 73 A store can be rented for R300 a month payable in advance. If the lessee pays 3 months in advance, the owner will allow interest at 7 % converted monthly. What is the size of the payment at the beginning of each 3 months that is equivalent to R300 at the beginning of each month?

Exercise 74 On 1 June 1977 a couple opened a savings account for their daughter with a deposit of R25 in a bank paying 5 % converted semi-annually. If they continue to make semi-annual deposits of R25 until 1 December 1984, how much will be in the account on 1 June 1985?

Exercise 75 Instead of taking R5 000 from an inheritance, a person decides to take 60 monthly payments with the first payment to be made immediately. If interest is allowed at 6 % converted monthly, what will be the size of each payment?

Exercise 76 A debt of R5 000 with interest at 4 % compounded quarterly is to be discharged by 8 equal quarterly payments, the first due today. Determine the quarterly payment.

Exercise 77 A corporation sets aside R10 000 at the beginning of each year to create a fund in case of future expansion. If the fund earns 3 %, how much does it amount to at the end of the tenth tear?

Exercise 78 Instead of paying R125 rent at the beginning of each month for the next 8 years, M decides to buy a house. Considering money at 15 % compounded monthly, what is the cash equivalent of the 8 years' rent?

INTEREST CALCULATIONS

Exercise 79   Determine the value on 1 June 1980 of a series of payments of R425 every 6 months, if the first of these payments is to be made on 1 December 1983 and the last on 1 June 1990. Use an interest rate of 10 % compounded semi-annually.

Exercise 80   Determine the value on 1 September 1978 of a series of R80 monthly payments, the first of which will be made on 1 September 1982, and the last on 1 December 1984 if money is worth 6 % compounded monthly.

Exercise 81   What sum put aside on a boy's twelfth birthday will provide 4 annual payments of R2 000 for college expenses if the first payment is to be made on his eighteenth birthday? The rate is 7 % compounded annually.

Exercise 82   What sum of money should be set aside today to provide an income of R150 a month for a period of 5 years if the first payment is to be made 4 years hence and money is worth 6 % compounded monthly?

Exercise 83   On his wife's 59th birthday Mr Hubby made provision for her to receive R150 a month for 5 years with the first payment to be made on her 65th birthday. If the investment earns 7 % converted monthly, how much money must be set aside?

Exercise 84   On 1 June 1978 a woman deposits R3 000 in an investment paying 5 % compounded annually. On June 1, 1984, she makes the first of 4 equal annual withdrawals from the account. Determine the size of the withdrawals if the account is closed with the last of these withdrawals?

Exercise 85   On 1 June 1977 a minor receives an inheritance of R6 500. This money is placed in a trust fund earning 7 % compounded semi-annually. On 1 December 1982 the child will turn 18 and be paid the first of 8 equal semi-annual payments from the fund. How much will each payment be?

APPENDICES

APPENDICES

APPENDIX 1
Revision Exercises

Exercise 1 Mike earned the following for selling magazines on a sample of 8 different days:
R8; R5; R11; R6; R4; R6; R7; and R9.
Assume that he worked for 6 months with 20 working days per month.

Calculate:

1.1 The mean
1.2 The median
1.3 The mode
1.4 The quartile deviation
1.5 The middle 70 % range
1.6 What can Mike say with confidence of 99 % about the mean for the six month period?
1.7 If the average number of magazines sold per salesman is 8, is there a significant difference at a 1 % significance level between Mike's average sales and the average of all the salesmen?

Exercise 2 A new operator in a manufacturing plant requires one month's training before being able to carry out an assembly operation effectively. After completion of this period of training an operator should be able to assemble a complete unit in 36 min on the average. This is the time allowed for in production calculations. A group of 9 new operators recorded the following times after they had completed their training:
35; 30; 29; 28; 34; 36; 27; 32; 37 (all in minutes)

2.1 Carry out a test of significance at a 1 % level to find whether this group of operators indicates that the time of 36 minutes could be revised downwards for all new operators.
2.2 Estimate at a 98 % level of confidence the average time required by all new operators to assemble a complete unit.
2.3 Determine the quartile deviation of this distribution.
2.4 Determine the mean absolute deviation of this distribution.

Exercise 3 The following lists are of times taken by employees of two different banks to perform the same task. The times are expressed in minutes.

	Bank A				Bank B		
6	2	5	9	7	8	5	5
4	5	4	4	11	8	9	8
5	7	5	8	9	7	5	8
6	6	4	3	4	7	8	6

3.1 For each set of employees, calculate a relative measure of dispersion and use your results to decide if the one set of workmen is better than the other.

3.2 Calculate Pearson's Second Coefficient of skewness for each distribution and compare the two banks using your results.

3.3 Is the mean time taken by the employees of bank B significantly higher than the mean time taken by the employees of bank A? Conduct the test at a significance level of 5 %.

Exercise 4 It has been established from past records that the average number of matches packed into boxes is 52. After a slight change in the packing procedure, a sample of 10 boxes were found to have the following contents:
48; 49; 50; 50; 51; 50; 52; 53; 54; 50

4.1 Compute the mean and standard deviation of the number of matches per box.

4.2 Discuss whether this sample shows evidence of a real change in the average number of matches being packed by making use of a 0.05 significance level.

4.3 Determine the middle 90 % range of the distribution of matches.

Exercise 5 The ages (in years) of a sample of 10 part-time students are:
19; 41; 23; 36; 17; 22; 51; 26; 34; 31

5.1 Calculate the mean absolute deviation.

5.2 Which value occupies the middle position?

5.3 Determine the middle 80 % range.

5.4 Construct a 98 % confidence interval estimate for the mean ages of all the part-time students.

Exercise 6 The average retail price for oranges in 1984 was 38.5 cents per kilogram. Recently a sample of 15 markets reported the following prices for oranges in cents per kilogram:

REVISION EXERCISES

37.9   44.2   44.5   38.2   42.4
43.0   40.0   42.6   40.2   37.5
44.1   45.2   41.8   35.6   34.6

6.1 Determine the range.
6.2 Determine the sample mean and standard deviation.
6.3 Is there a significant change in the average price of oranges since 1984 at a 0.02 level of significance?
6.4 What is the probability that the mean sample price will be within two cents of the 1984 mean price?
6.5 Construct a 98 % confidence interval estimate for the recent mean price for all oranges.

Exercise 7 From the following table, which shows the number of workmen employed by 50 firms, construct:
7.1 a frequency distribution table;
7.2 a histogram;
7.3 a polygon; and
7.4 ogives.

No. of workmen:	120–124	125–129	130–134	135–139	140–144	145–149
No. of firms:	1	2	6	26	13	2

Exercise 8 The following table sets out the performance of a machine which manufactures rugby balls, when it was new and five years later:

Life of balls (in hours)	Machine new	Machine old
0–299	2	9
300–599	5	25
600–899	23	67
900–1 199	63	78
1 200–1 499	70	35
1 500–1 799	56	16
1 800–2 099	11	0
	230	230

8.1 Calculate Pearson's Second coefficient of skewness for each distribution.
8.2 Compare the shapes of the distributions using appropriate graphs on the same axis. Comment on each one.
8.3 Use the information given to decide whether the machine is due for an overhaul. That will be the case if the results obtained from

the old machine are significantly lower than that of the new machine. Use = 0.05.

8.4 The manufacturer set the control limits on the machine's performance when it was new, and made provision for the replacement of all balls with a life of less than 900 hours. What percentage of the balls does he expect to replace when the machine
— is new;
— is five years old?

Exercise 9 In a company there are 500 employees: 300 men and 200 women. The men average 1.8 m in height with a standard deviation of 0.08 m. The women average 1.7 m in height with a standard deviation of 0.08 m.

9.1 What is the average height of all 500 people?
9.2 How many women will be over 1.8 m in height?
9.3 What interval of heights contains the middle 50 % of the men?
9.4 What is the 75th percentile of the women's heights?

Exercise 10 In a statistics exam, the average mark was 55 % with a standard deviation of 18 %. If the lecturer wishes to control the percentage of students receiving an A to the top 15 % of the class, what should his cut-off mark be?

Exercise 11 The following is a table of scores obtained on an employment aptitude test by 1 000 applicants.

Interval	Frequency
92–100	60
83–91	140
74–82	160
65–73	120
56–64	140
47–55	80
38–46	119
29–37	81
20–28	50
11–19	32
2–10	18

11.1 Calculate the mean, median, mode and standard deviation of the distribution.
11.2 Discuss the shape of the distribution based on the calculations and a rough sketch.

REVISION EXERCISES

Exercise 12 Data have been collected on the life of two brands of light bulbs as follows:

$\overline{X}_1 = 800$ hours

$\sigma_1 = 100$ hours

$n_1 = 80$

$\overline{X}_2 = 770$ hours

$\sigma_2 = 60$ hours

$n_2 = 60$

12.1 Calculate the coefficient of variation and interpret the result.

12.2 Is the life time of the second brand significantly lower than that of the first brand? Use = 0.05.

Exercise 13 Your company manufactures springs. The lifetimes of a sample of 100 springs were recorded as follows:

Hours	Number of springs
2 000 to 3 000	3
3 000 to 4 000	13
4 000 to 5 000	19
5 000 to 6 000	27
6 000 to 7 000	21
7 000 to 8 000	12
8 000 to 9 000	5

13.1 Calculate the mean and standard deviation of the life-times of these 100 springs.

13.2 Determine Pearson's Second coefficient of skewness and interpret your answer by making use of a rough sketch.

13.3 Estimate at a 90 % level of confidence the average life-time of the springs.

13.4 Find the probability that a spring selected at random will have a life-time of between 4 700 and 6 300 hours.

13.5 Approximately 20 % of the springs have a life-times of less than the guaranteed minimum hours. What is the minimum number of hours?

Exercise 14 Items are manufactured to the same length on two machines. The results of test on items from each machine are given below:

Class interval (cm)	Machine a	Machine b
19–21	4	12
21–23	6	28
23–25	20	30
25–27	50	18
27–29	14	12
29–31	6	—

14.1 Compare the shapes of the two distributions using appropriate graphs.

14.2 Calculate Pearson's Second coefficient of skewness for each distribution and interpret your results.

14.3 Determine the inter-quartile range for each distribution.

14.4 Determine the modal value graphically.

14.5 Determine the median value graphically.

14.6 Do you think the machines are producing items of the same average lengths? Use = 0.05

14.7 Items shorter than 24 cm are to be rejected. What percentage of the items do you expect to reject for:
— Machine A?
— Machine B?

Exercise 15 The production division of Continental Motors planned a campaign to improve plant safety. In preparation, the following accident data were compiled for a sample of 50 weeks:

Number of accidents	Number of weeks
0–4	6
5–9	25
10–14	11
15–19	7
20–24	1

15.1 Determine the total number of accidents for the 50 weeks.

15.2 Determine the modal number of accidents per week using the formula and estimate the modal number of accidents using an appropriate graph.

15.3 Determine Pearson's Second coefficient of skewness and interpret your answer using a rough sketch.

15.4 What percentage of weeks has more than 12 accidents? Use an appropriate graph to answer the question.

15.5 Estimate a 99 % confidence interval estimate for the mean number of accidents.

15.6 If the average number of accidents was 10.5 last year, does the current parameter show a significant decrease?

15.7 What is the probability that in a randomly selected week, there will be no more than 1 accident? In how many weeks do you expect no more than 1 accident?

15.8 What is the probability that in a randomly selected week there will be between 4 and 6 accidents? In how many weeks do you expect between 4 and 6 accidents?

15.9 If bonuses were paid on a basis of to the 10 % of the weeks with the lowest accident rate. What is the maximum accident rate that will be acceptable before bonuses can be paid?

Exercise 16 Tiny Smith, a production supervisor, is worried about an elderly employee's ability to keep up the minimum work pace. In addition to the normal work breaks, this employee stops for short rest periods an average of 2 times per half an hour. The rest period is a fairly consistent 3 minutes each time. Tiny has decided that if the probability of the employee's resting for 6 minutes or more per hour is greater than 0.8, he will move the employee to a different job. Should he do so?

Exercise 17 A quality control scheme takes a sample of 10 items and checks the whole batch if more than one is defective. Given that the probability of a defective item is 0.05, what is the probability that the whole batch will be checked?

Exercise 18 A retail store manager has determined that in a sample of 100 customers 60 % are female. Construct a 98 % confidence interval estimate for the true proportion of customers that are female.

Exercise 19 The manufacturer of a certain oil-additive claims that the mean net mass of jars of his product is 1 kg. A random sample of 49 of a large consignment supplied to your company is found to have a mean mass of 0.98 kg with a standard deviation of 0.02 kg. Test the manufacturers claim at the 0.05 significance level.

Exercise 20 An assembly line has 8 grinders. At least 6 of these must be in operation in order to maintain adequate material for the assembly line to produce on schedule. The probability that a grinder will be inoperative on any day because of mechanical failure is 0.15. There is also the probability of 0.10 that any operator will be absent. Any operator can use any machine.

What is the probability that the production line will not meet its schedule on any particular day because of the grinder section?

Exercise 21 A high-speed machine produces polyethylene sheeting. It produces an average of 3 flaws in each 100 m of the material. If the inspector finds 4 or more flaws in a 100 m piece, he will stop the machine.

21.1 What is the probability that the machine will be stopped upon inspection of a 100 m piece?

21.2 What is the probability that a 50 m piece will contain less than 2 flaws?

21.3 What is the probability of producing at least 7 but not as many as 11 flaws in a 200 m piece?

Exercise 22 Assume that your firm wants to test whether the average output of a new machine is at least 400 units per hour as claimed. A random sample of 64 production hours is selected, and the sample mean is found to be 395 units with a standard deviation of 10. At a 98 % confidence level test the appropriate hypothesis.

Exercise 23 Trucks stop at a highway weigh station at an average of 6 per 8-hour day:

23.1 If the operator goes to lunch for 1 hour, what is the probability that no trucks will arrive during that period?

23.2 If he takes a $1\frac{1}{2}$-hour lunch period, what is the probability that 2 or more trucks will be missed?

Exercise 24 In a meat packing plant, machines X and Y account for 60 % and 40 % of the plant's output, respectively. It is known that 1 % of the packages from machine X and 2 % of those from machine Y are improperly sealed.

24.1 If a package is selected at random, what is the probability that it will be improperly sealed?

24.2 If a package selected at random is improperly sealed, what is the probability that it came from machine X?

Exercise 25 In an assessment of job performance the marks awarded are normally distributed with a mean of 55 and a standard deviation of 11.

25.1 In a group of 300 employees, how many would you expect to obtain over 75 marks?

25.2 From past performances it can be seen that approximately 9 % of employees obtain unsatisfactory marks. What is the minimum satisfactory mark?

25.3 What portion of the marks will fall between 45 and 50?

REVISION EXERCISES

Exercise 26 A chicken breeder claims that the average mass of a particular group of chickens exceeds 1.5 kg. Before agreeing to a purchase, a particular customer selected a sample of 25 chickens which yielded a sample mean of 1.6 kg and a standard deviation of 0.1 kg. Should this claim be rejected at the 0.05 level of significance?

Exercise 27 Thirty percent of Uncle Joe Beer is produced at the company's old installation and the rest at the new plant. Representatives of the buyers have been invited to taste-test 10 bottles of Uncle Joe Beer. What are the chances that:
    27.1 All of them are produced at the new plant?
    27.2 At least two will be from the old plant?
    27.3 Half of them are produced at the new plant and half at the old plant?

Exercise 28 In a survey of customers' opinions, random samples of 50 customers were questioned at two stores. The results of the survey showed that 66 % of the customers were satisfied with the service in store A compared with 54 % in store B. Use a 1 % level of significance to test whether there is a significant difference between the proportion of satisfied customers in the two stores.

Exercise 29 Experience has shown that, on the average, 2 % of an airline's flights suffer a minor equipment failure in an aircraft. Estimate the probability that the number of minor equipment failure in the next 50 flights will be:
    29.1 Zero
    29.2 At least two
    (Note: $\mu = n\pi$)

Exercise 30 A company mass-produces electronic calculators. From past experience it is known that 90 % of the calculators will be in working order if the production process is working satisfactorily. An inspector randomly selects 5 calculators from the production line every hour and carries out a rigorous check. What is the probability that the sample will contain at least two defective calculators?

Exercise 31 Manufactured items are sold in boxes which are stated to contain a mass of at least 40 kg. The actual mass in a box varies, being approximately normally distributed with a mean of 41.2 kg and a standard deviation of 0.8 kg.
    31.1 Calculate the proportion of boxes whose mass is between 40 kg and 42 kg.

31.2 Calculate the mass below which 20 % of the lightest boxes fall.
31.3 All boxes containing less than 40 kg are scrapped at a cost of R100 per box. Calculate the scrapping cost associated with the packing of 50 boxes.
31.4 To what mean mass should the box contents be adjusted, the standard deviation unchanged, if only 1 % of the boxes are to be scrapped?

Exercise 32 If an average of 40 service calls are required during a typical eight-hour shift in a manufacturing plant, what is the probability that no service calls will be required during a particular hour?

Exercise 33 In a factory components are received from two suppliers, A and B. Using a quality control inspection scheme, random samples of 100 components from each supplier were taken. These showed that 10 % of the components from supplier A were defective and that 6 % of those from supplier B were defective. Apply an appropriate statistical test of significance to see if there is strong evidence that supplier B's quality is better than that of supplier A.
Use $\alpha = 0.05$.

Exercise 34 A new word-processing unit is fully warranted for one year. The average number of warranty service calls for a unit during its first year is four. What is the probability of exactly 5 warranty service calls for a unit during the first year?

Exercise 35 The tear strength of a particular paper product is known to be normally distributed. If a random sample of 9 rolls yielded a mean tear strength of 225 kg/m^2 with a standard deviation of 15 kg/m^2, construct a 90 % confidence interval estimate for the average tear strength.

Exercise 36 A random sample of 400 television tubes was tested and 40 tubes were found to be defective. With a confidence coefficient of 90 %, estimate the interval within which the the true fraction defectives lies.

Exercise 37 Twenty containers of a commercial solvent randomly selected from a large production lot have a mean mass of 24.5 kg. with a standard deviation of 0.3 kg. What can we say with 95 % confidence about the true mean mass of a container of this solvent?

Exercise 38 Given that 35 one-liter cans of a certain kind of paint covered on the average 45.22 m^2 with a standard deviation of 2.28 m^2, construct a 99 % confidence interval estimate for μ.

Exercise 39 To compare two kinds of bumper guards, six of each kind were mounted on 12 cars. Then each car was run into a concrete wall at 10 km/h. The following results were obtained for the repair costs in rand:
$X_1 = 87.6$
$s_1 = 12.54$
$X_2 = 107.33$
$s_2 = 11.28$
Use a significance level of 1 % to test whether the difference between the mean repair costs is significant.

Exercise 40 A sample study was made of the number of business lunches that executives claim as deductible expenses per month. If 60 executives in the insurance industry averaged 9.6 such deductions with a standard deviation of 1.8 in a given month, while 50 bank executives averaged 8.4 with a standard deviation of 2.1, test at a 0.01 level of significance whether the difference between these two sample means is significant.

Exercise 41 According to specifications, the mean time required to inflate a rubber raft is 7.5 seconds. As it it has been suggested that this figure might be too low, a random sample of 45 of the rafts are inflated, yielding $\overline{X} = 7.6$ seconds and $s = 0.6$ seconds. What can we conclude at the 0.01 level of significance?

Exercise 42 The security department of a warehouse wants to know whether the average time required by the night watchman to walk his round is 12.0 minutes. If, in a random sample of 26 rounds, the watchman averaged 12.3 minutes with a standard deviation of 1.2 minutes, what can we conclude about the average time?

Exercise 43 Among 80 fish caught in a certain lake, 28 were inedible as a result of the chemical pollution of their environment. Construct a 95 % confidence interval for the true proportion of fish in this lake which are inedible as a result of the chemical pollution.

Exercise 44 The manufacturer of a spot remover claims that her product removes at least 90 % of all spots. If in a random sample, the spot remover removes only 10 of 14 spots, test this claim at the 0.05 level of significance.

Exercise 45 An agricultural cooperative claims that 95 % of the watermelons that are shipped out are ripe and ready to eat. Find the probability that among 10 watermelons that are shipped out:
    45.1   At least 8 are ripe and ready to eat.

45.2 At most nine are ripe and ready to eat.

45.3 Anywhere from seven to nine are ripe and ready to eat.

Exercise 46 If the number of blossoms on a rare cactus is on the average 2.4, what are the probabilities that such a cactus will have:

46.1 no blossoms;

46.2 at least one blossom;

46.3 more than 3 blossoms?

Exercise 47 The given table shows the advertising expenditure and resulting sales of a company over seven years:

Year	Advertising (R'000)	Sales (R'000)
1980	1.9	10
1981	2.4	12
1982	2.3	13
1983	3.7	17
1984	2.9	14
1985	2.6	13
1986	3.1	15

47.1 Forecast advertising expenditure for 1987 using the method of least squares.

47.2 Calculate the degree of correlation between advertising and sales. Comment on your result using the coefficient of determination.

47.3 Is the correlation coefficient significant at a 1 % level?

47.4 Find the regression equation in order to estimate sales for 1987 within a confidence level of 99 %.

47.5 Determine the average increase ratio for sales over the seven years, using an appropriate measure of central tendency.

Exercise 48 The following table shows the number of car and truck registrations in a certain town and the sales of tyres by J. Jenk and Son for the same period.

Year	Registrations	Tyres sold
1981	6 000	13 000
1982	6 400	12 000
1983	7 300	14 000
1984	7 500	14 500
1985	8 000	16 000

REVISION EXERCISES

48.1 Predict the number of registrations for 1986 using the method of least squares.

48.2 What was the average increase in the number of registrations over the period?

48.3 Calculate the coefficient of correlation between registrations and numbers of tyres sold and comment upon the result using the coefficient of determination.

48.4 Comment on the significance of the correlation coefficient at a 1 % level of significance.

48.5 Find the least squares regression equation of tyres on registration and predict the number of tyres you expect to sell for the estimated number of registrations for 1986 in 48.1.

Exercise 49 When buying almost any item, it is often advantageous to buy in as large a quantity as possible. The unit price is usually less for the larger quantities. The data shown in the table were obtained to test this theory.

Number of units:	1	3	5	10	15
Cost per unit:	55	52	48	32	25

49.1 Can we conclude that the number of units ordered and the cost per unit are significantly correlated?

49.2 Find the 98 % confidence interval estimate for the price if you buy 17 items.

Exercise 50 The number of machine hours and maintenance cost (in rands) of six machines in a factory is shown in the following table:

Machine hours:	55	91	140	205	289	386
Maintenance cost:	19	29	41	52	59	69

50.1 What percentage of the variation in maintenance cost is explained by changes in machine hours?

50.2 Test the reliability of the correlation coefficient. Use $\alpha = 0.05$)

50.3 Calculate the expected maintenance cost after 300 machine hours within a confidence level of 95 %.

Exercise 51 To find the best arrangement of instruments on a control panel in a plant, three different arrangements were tested by simulating an emergency and observing the reaction time of the operators to correct the emergency. The reaction time of 12 operators (randomly assign to the different arrangements) are as follows:

Arrangement		Reaction time		
A	8	14	11	11
B	17	10	15	18
C	13	7	12	8

Determine whether the difference between the sample means can be attributed to chance. Use $\alpha = 0.05$

Exercise 52 The following frequency distribution was obtained in an aptitude test:

Marks	Frequency
4–7	4
8–11	14
12–15	44
16–19	29
20–23	9

Test on a 5 % level of significance whether these values come from a normal distribution.

Exercise 53 A researcher wishes to compare the mean writing lifetimes of five brands of similarly priced felt-tip markers. Tests conducted on samples of 10 markers of each brand revealed the following results which have been coded for computational eases.

Brand					Life					
A	3	9	7	0	3	2	1	3	5	7
B	1	7	9	5	4	6	8	9	3	4
C	2	3	1	0	3	2	3	0	3	4
D	9	9	7	9	5	6	6	5	8	9
E	5	6	9	10	9	8	5	9	6	9

Can one conclude from these data that the five brands differ with respect to mean writing lifetimes? Use $\alpha = 0.05$

Exercise 54 The number of typist errors per page of a 2 000 page manuscript is as follows:

No. of errors:	0	1	2	3	4
Observed no. of pages:	1 102	657	193	35	13

Fit a Poisson distribution to these data and use a χ^2-test to determine the goodness-of-fit at a 0.01 level of significance.

REVISION EXERCISES

Exercise 55 The number of deliveries per day arriving at a warehouse over a period of 50 working days is given in the table below.

No. of deliveries:	0	1	2	3	4 or more
No. of days:	11	18	10	7	4

Use the χ^2 test at the 5 % level to investigate whether the number of deliveries conforms to the Poisson distribution.

Exercise 56 A quality control engineer takes a daily sample of 5 tyres coming off an assembly line to check the numbers of tyres with imperfections. He wants to check on the basis of the following data, if the samples follow a binomial distribution at a 0.05 level of significance.

No. of imperfections:	0	1	2	3	4	5
No. of days:	2	6	5	4	2	1

Exercise 57 You are a manufacturer of computer devices and have three suppliers of computer chips. The following table shows the number of defective chips per order:

Order	Supplier A	Supplier B	Supplier C
1	7	3	5
2	5	4	8
3	6	5	8
4	4	5	7

Would you conclude that there is no real difference in the average number of defective products? Use the 0.05 risk level?

Exercise 58

Weekly Wage	Number of Employees
80–100	10
100–120	25
120–140	45
140–160	10
160–180	5
180–200	5

Test the assumption that the wages in this company conform to the Normal distribution using a 5 % significance level.

Exercise 59 The table below shows the quarterly sales of magazines. Do a seasonalized forecast per quarter for 1988. The figures are given in thousands of copies.

	1984	1985	1986	1987
1st	112	132	144	160
2nd	145	161	169	181
3rd	164	180	196	208
4th	123	135	143	155

Exercise 60 The accountant of a company has calculated the following unadjusted seasonal indexes based on the quarterly value of sales over a number of years:

Quarter	Index
September	80
December	110
March	90
June	80

The secular trend equation of the company is as follows:
$Y_c = 50 + 12X$
With:
- origin December 31, 1980
- X unit 6 months
- Y unit annual sales

60.1 If 1985 was the last year of data, how many years of data were used to determine the trend equation?

60.2 Project the seasonalized sales per quarter for 1986.

Exercise 61 A study is conducted to examine the relationship between the amount spent on advertising a new product and consumer awareness of the product based on the proportion of people who have heard of it. Suppose a sample shows the following data for four different products:

Advertising Expenses (R)	Consumer Awareness (%)
800	90
180	20
100	10
200	50

61.1 Predict consumer awareness if R500 is spent on advertising.

61.2 What percentage of the variation in consumer awareness is explained by changes in amount spent on advertising?

Exercise 62 The table below shows the staff turnover experienced by a company over a four-year period. Forecast a seasonalized quarterly turnover for 1989.

Year	I	II	III	IV
1985	3.1	2.6	4.8	4.3
1986	3.7	2.8	5.2	4.7
1987	4.1	3.4	5.8	4.9
1988	4.5	4.0	6.0	5.3

Exercise 63 An assistant secretary in the commerce department has the following data (R'000 000) describing the value of grain exported during the last four years.

Year	I	II	III	IV
1981	1	3	6	4
1982	2	2	7	5
1983	2	4	8	5
1984	1	3	8	6

63.1 Estimate the seasonalized grain exports for the second and third quarters of 1985.

63.2 Deseasonalize the first and forth quarters grain exports for 1984.

Exercise 64 The East branch of the Woodpecker Timber outlet has recorded sales for three of its outlets:

Sales (R'000)

	1985	1986
Pine tree	84.7	86.3
Elmwood	72.6	70.9
Oak tree	64.5	87.6

With 1985 = 100, compute the unweighted average of relatives price index for 1986.

Exercise 65 The analysis of products imported by a certain company was as follows:

	Price per unit (R)		Number of units	
Product	1975	1985	1975	1985
X	62	74	17	27
Y	57	52	12	17
Z	37	47	20	26

Calculate for 1985 with 1975 as base year:
65.1 Laspeyres price index
65.2 Paasche's volume index

Exercise 66

	1987		1988	
Item	Price (R)	Quantity	Price(R)	Quantity
A	1.5	20	1.6	20
B	0.8	40	0.85	60
C	0.4	60	0.55	50
D	2.0	100	2.55	90

66.1 Calculate a current year weighted index number showing the overall change in price.
66.2 Calculate a base year weighted index showing the overall change in volume.

Exercise 67 The table below shows the the prices and annual consumption of the raw materials used in a particular brewery in 1987 and 1988.

	1987		1988	
Material	Price per ton (R)	Consumption	Price per ton (R)	Consumption
Malt	49	19 874	46	25 116
Hops	512	732	724	496
Sugar	46	1 865	51	2 486
Wheat flour	31	873	27	2 093

67.1 Calculate a current weighted index number showing the overall change in raw material prices.
67.2 Calculate Irving Fisher's Ideal Price index.

Exercise 68 Tixif Limited sells three types of chain-saws. Company records showed the prices and quantities sold as follows:

	1985		1987	
Type	Price (R)	Number sold	Price (R)	Number sold
X	30	22	40	30
Y	50	31	60	40
Z	120	8	99	12

The owner of Tixif wants to know whether or not the average price paid for his chain-saws has increased or decreased from 1985 to 1987 and by how much. What would you tell him if you use Irving Fisher's Ideal price index?

Exercise 69 A deposit of R10 000 and instalments of R600 at the end of each month for 20 years buy a house. If interest is calculated at 18 % per year compounded monthly, what is the cash price of the house?

Exercise 70 The rent of a building is R1 500 per year payable in advance. If the interest rate is 6 % per year compounded monthly, what will the equivalent monthly rental, payable in advance, be?

Exercise 71 An amount of R200 was deposited at a building society for four years at 17 % per year. The accumulated amount was then withdrawn and deposited at a bank for 6 years where interested was compounded quarterly at 18 % per year. The total amount of the investment after 10 years was then withdrawn and loaned to a student for 5 years at 20 % per year compounded annually. How much must the student pay the lender to settle the loan in full after these five years?

Exercise 72 Mr T. Bone took out a R100 000 loan on a steak-house over a 10 year period at an interest rate of 12 % per annum compounded monthly. After 3.5 years interest rates climbed to 15 % per annum compounded monthly.
    72.1   If his repayments were at the end of each month, how much did Mr Bone owe at the end of the first 3.5 years?
    72.2   What was his monthly repayments for the remaining 6.5 years?

Exercise 73 A person deposits R15 at the end of each month for a period of four years. Thereafter, because of unpaid leave, he cannot make deposits for 1 year. At the end of the year he resumes his payments on the old scale and continues for another two years. What is the final amount for the 7 year period under review if interest was calculated at 12 % per year compounded monthly?

Exercise 74   Mr D. Cute bought a new car paying R2 000 deposit and R200 at the beginning of each month for 4 years. If current interest rates are 9 % per annum compounded monthly, what was the cash price of the car?

Exercise 75   A lounge suite is bought on hire-purchase on the following conditions:
- 12 % is added to the cash price of R250.
- This balance is to be repaid with monthly installments at the beginning of each month for 12 months.

What are the monthly payments if interest is 18 % per year compounded monthly?

Exercise 76   Mr B. Ware deposited R15 000 at 9 % per year compounded quarterly in a finance company. Three years later he deposited a further R6 000 in the account for another 2 years at the same interest rate. How much will Mr Ware earn in interest at the end of the 5 years?

Exercise 77   Mr B. Dazzle has been offered two fairly attractive contracts by opposing publishers for the rights to his impressive autobiography. Lystin and Dumpin has offered him R100 000 outright, while Uppiti Dumpity offered royalties of R13 000 every year-end for the next 15 years. If the current return is 10 % per year, compounded annually, which of the contracts should be accepted? Assume that if Mr Dazzle accepted the R100 000 outright, he would deposit it in a 10 % per year interest bearing account.

Exercise 78   A student wants to save R1 000 for a trip after graduation 4 years from now. How much must she invest at the beginning of each month starting now if she gets 18 % compounded monthly on her savings?

Exercise 79   What sum of money was invested 2 years and 10 months ago if R1 000 is accumulated at 14 % interest compounded quarterly?

Exercise 80   Des and Mel Grimm had R8 000 and R9 000 respectively in savings. Des deposited his funds in a fixed-term deposit for 3 years at 12 % compounded quarterly. Mel found a return of 16 % p.a. compounded quarterly from a less reputable financial institution. After 3 years they pooled the accumulated value of their respective deposits and deposited the money with an institution offering 12 % p.a. compounded monthly for 5 years. How much did the Grimm brothers have in total at the end of the 8 year period?

Exercise 81   Mr C. Daar wants to spend 5 years researching the timber industry. He calculated that he needs R1 200 per month to live on. How much must he deposits today in an account earning 12 % p.a. compounded monthly in order to withdraw R1 200 at the beginning of each month over the next 5 years?

Exercise 82 Find the amount of R3 500 due in 4 years and 6 months if money is worth 8 % compounded yearly.

Exercise 83 Find the deposit paid on a house offered for sale at R100 000 if the debt is to be completely paid off by monthly payments of R1 200 at the end of each month for 5 years. Interest is 18 % p.a. compounded monthly.

Exercise 84 A store can be rented for R350 a month payable in advance. The owner will accept a single cash payment at the beginning of a year for a year's lease. If interest is 18 % compounded monthly, what would be the cost of a year's lease?

Exercise 85 Mr Smit invests R4 000 at 12 % p.a. compounded quarterly for 62 months. What amount will be received after the period?

APPENDIX 2
Basic Mathematics for Statistics Students

This textbook was written for students who need a working knowledge of statistics, but do not have a strong mathematics background. Since statistics requires the use of many formulas, it is advisable to read and review this section thoroughly and not to proceed until you have mastered the material.

The objective of these notes is to review basic operations of fractions, decimals, signed numbers, exponents, logarithms, factorials, algebraic expressions, etc.

1 Fractions An arithmetic fraction is the division of one number by another. In the fraction $\frac{7}{8}$, 7 will be referred to as the numerator and 8 will be referred to as the denominator.

Rules:

- Any fraction in which the denominator equals the numerator has a value of 1
  $$\frac{8}{8} = 1$$

- Division into zero (i.e. zero is the numerator) is a mathematically undefined operation, the fraction always equals zero.
  $$\frac{0}{8} = 0$$

- When multiplying or dividing both the numerator and the denominator by the same value, the fraction value does not change.
  $$\frac{3}{8} \times \frac{2}{2} = \frac{3}{8}$$

- When adding or subtracting fractions, the denominators of all the fractions must be the same. The numerators are then added or subtracted, and the result then divided by the common denominator.

$$\frac{3}{8} + \frac{2}{8} = \frac{5}{8}$$

$$\frac{6}{8} - \frac{2}{8} = \frac{4}{8}$$

If the denominators are different, a common denominator is found by multiplying the denominators.

$$\frac{1}{2} + \frac{3}{8}$$

Common denominator: $2 \times 8 = 16$.

$$\frac{1}{2} = \frac{8}{16} \text{ (both numerator and denominator are multiplied by 8)}$$

$$\frac{3}{8} = \frac{6}{16} \text{ (both numerator and denominator are multiplied by 2)}$$

$$\frac{8}{16} + \frac{6}{16} = \frac{14}{16}$$

❑ To multiply fractions, multiply both numerators and denominators.

$$\frac{1}{2} \times \frac{3}{8} = \frac{3}{16}$$

❑ To multiply fractions by a whole number, multiply the numerator by the whole number, maintaining the same denominator.

$$\frac{1}{2} \times 3 = \frac{3}{2}$$

To multiply mixed numbers (i.e. a whole number and a fraction), multiply the whole number by the denominator and add the numerator to that result. This is now the new numerator with the denominator remaining the same as in the original mixed number.

$$2\frac{1}{3} \times 1\frac{1}{2} = \frac{(2 \times 3 + 1)}{3} \times \frac{(1 \times 2 + 1)}{2}$$

$$= (7 \div 3) \times (3 \div 2)$$

$$= 21 \div 6$$

$$= 3.5$$

❑ When dividing by a fraction, inverse the fraction and then multiply.

$$\frac{2}{3} \div \frac{3}{8} = \frac{2}{3} \times \frac{8}{3}$$

$$= \frac{16}{9}$$

BASIC MATHEMATICS FOR STATISTICS STUDENTS

❏ Always reduce a fraction to its lowest terms by dividing both the numerator and denominator by a common factor.

$$\frac{6}{12} = \frac{6 \div 6}{12 \div 6} = \frac{1}{2}$$

$$\frac{8}{12} = \frac{8 \div 4}{12 \div 4} = \frac{2}{3}$$

2 Decimals A decimals can be thought of as a fractions, the numerator of which has been divided by the denominator.

$$\frac{3}{8} = 3 \div 8 = 0.375$$

Rounding off

When a result has more decimal places than necessary, it requires rounding off to the appropriate number of decimal places.

❏ If the number subsequent to the required cut-off number is greater than 5, increase the number to its left by one. If it is less than 5, disregard all the numbers to the right of the cut-off digit.

0.089 to the nearest hundredth = 0.09
0.083 to the nearest hundredth = 0.08

❏ In the case of a number ending with a decimal half, round the cut-off digit to the nearest even number.

2.185 to the nearest hundredth = 2.18
2.175 to the nearest hundredth = 2.18

3 Signed numbers Rules

❏ When adding numbers of the same sign, find the sum of the numbers and use the sign common to all factors.

+2 +3 +4 = +9
(−2) + (−3) + (−4) = −9

❏ When adding numbers of different signs, find the sum of the positive numbers and the sum of the negative numbers, and then subtract the smaller sum from the bigger one and designate the sign of the bigger sum.

(−2) + 3 + 4 + (−1) = +7 − 3 = +4
2 + (−3) + (−4) +1 = +3 − 7 = −4

❏ When subtracting a negative number, change the sign of the negative number being subtracted and add the number to the rest.

2 + 3 − (−4) = 2 + 3 + 4 = +9

- When multiplying or dividing by the same sign, the answer will always be positive.

  $-2 \times -2 = +4$

  $2 \times 2 = +4$

  $\dfrac{-2}{-4} = \dfrac{1}{2}$

- When multiplying or dividing unlike signs, the answer is always negative.

  $-2 \times 2 = -4$

  $3 \times -4 = -12$

  $\dfrac{-6}{2} = -3$

4 Exponents

When a number is multiplied by itself one or more times, a superscript number can be placed at the upper right-hand side of the number.

$2 \times 2 = 2^2$

$2 \times 2 \times 2 = 2^3$

Rules

- Exponents with the same base number are multiplied by adding the exponents.

  $2^3 \times 2^2 = 2^5$

- Exponents with the same base number are divided by subtracting the exponents.

  $\dfrac{2^5}{2^3} = 2^2$

- A negative exponent is equivalent to the reciprocal of that factor.

  $2^{-5} = \dfrac{1}{2^5}$

5 Roots

- The square-root of a number is that quantity which, when multiplied by itself, equals the number.

  $16 = 4 \times 4$

  $\sqrt{16} = 4$

- All roots can also be written as fractional exponents.

  $\sqrt{16} = 16^{1/2}$

BASIC MATHEMATICS FOR STATISTICS STUDENTS

$$\sqrt[3]{16} = 16^{1/3}$$
$$\sqrt[4]{5} = 5^{1/4}$$

6 Factorial notation (!) This notation is a shorthand way to identify the product of a particular set of numbers. 5! (five factorial) stands for the product of all positive numbers starting with 5 and proceeding downward until 1 is reached. That is:

$$5! = 5 \times 4 \times 3 \times 2 \times 1 = 120$$

Notes

- ❏ The number in front of the factorial symbol will always be positive or 0.
- ❏ The last number in the series is always 1, except in the case of 0!. The value of zero factorial is defined to be one.

7 Summation notation (Σ) This Greek capital letter sigma (Σ) stands for the direction 'sum up the appropriate values'. $\sum_{i=1}^{10} x_i$ means the sum of all the values of x assigned to cases 1 through 10. This index system must be used whenever only part of the available information is to be used. In statistics, however, we will usually use all the available information, and to simplify the formulas we will make an adjustment. This adjustment is an agreement that allows us to do away with the index system in situations where all values are used.

$\sum x_i$ is the same as $\sum x$ if all the data are used.

If $x = 2, 4, 1, 3$ and 5

$$\sum x = 2 + 4 + 1 + 3 + 5$$
$$= 15$$
$$\sum x^2 = 2^2 + 4^2 + 1^2 + 3^2 + 5^2$$
$$= 15$$

8 Logarithms The logarithm of a number is the power to which a base number must be raised to produce that number.

Common logarithms

Although any positive number other than one can be used as the base for a system of logarithms, 10 is the best base for computational work. When the base is not indicated, 10 is understood.

$1\,000 = 10^3$

$\log 1\,000 = 3$

The anti-logarithm is the number that corresponds to a logarithm of that number. For example, 1 000 is the anti-logarithm of 3 because $1\,000 = 10^3 = \log 1\,000$

$$x = y^n$$
$$\log x = \log y^n$$
$$= n \log y$$
$$n = \frac{\log x}{\log y}$$

or $\quad x = y^n$
$$l_n x = l_n y^n$$
$$= n l_n y$$
$$n = \frac{l_n x}{l_n y}$$

Natural logarithms

Logarithms which use $e = 2.718 \ldots$ as the base, are called natural logarithms and can be denoted by $\ln x$. The anti-logarithm of the natural log $\ln$, is e^x.

$\ln 5 = 1.6094$

$e^{1.6094} = 5$

9 Percentages It is often vitally important to be able to compare two or more numbers. Of all the possible methods available to make such a comparison, the percentage is probably the easiest to work with.

❑ Percent indicates the relationship between the number preceding the percentage symbol and 100. The basis of comparison is always 100.

$65\,\% = 65/100$

❑ To change from percentages to decimals, move the decimal point two places to the left (that is divide by 100) and omit the percentage sign.

$65\,\% = 0.65$

❏ To change from a decimal to a percentage, move the decimal two places to the right (multiply by 100) and add the percentage sign.
0.235 = 23.5 %

The following are three common problems:

(1)
A number and percentage are given and you are required to find the missing link.
Find the value of 15 % of 60.
Method: Express the percentage as a fraction or decimal and multiply by the number.
$$\frac{15}{100} \times 60 = 9 \text{ or } 0.15 \times 60 = 9$$

(2)
The rate of one number compared with another needs to be determined.
Express 4 as a percentage of 10.
Method: Express the relationship as a fraction and multiply by 100.
$$\frac{4}{10} \times 100 = 40\%$$

(3)
Percentage increase and decrease
$$\text{Percentage increase} = \frac{\text{increase}}{\text{original value}} \times 100$$
$$\text{Percentage decrease} = \frac{\text{decrease}}{\text{original value}} \times 100$$

The basic wage was increased from R60 per week to R68 per week. Determine the percentage increase.
$$\text{Percentage increase} = \frac{8}{60} \times 100 = 13.3$$

or: The basis of comparison is always 100 and an answer more than 100 means an increase and less than 100 a decrease.
$$\text{Percentage increase} = \frac{\text{increase}}{\text{original value}} \times 100$$
$$= \frac{68}{60} \times 100$$
$$= 113.3$$

Increase is 13.3 %
The price of petrol changed from 98 c per liter to 87 c per liter, what was the percentage decrease?
$$\text{Decrease} = \frac{87}{98} \times 100 = 88.78$$
which means a decrease of 11.22 %

10. Formulas   The ability to use formulas plays a major role in the study of statistics. A formula is an algebraic equation that describes the procedure that must be followed in order to find a desired piece of numerical information.

Most of the formulas you will deal with require only the use of substitution. That is, the variables in the formula are substituted by specified values and evaluated — the result is relatively easy to obtain. If $P = 100$, $r = 5\%$ and $t = 5$, determine I if

$I = Prt$
$= 100 \times 0.05 \times 5$
$= 25$

Rules

❏ When solving an equation, like terms should be put on the same side of the equation.

❏ Any operation performed on one side of the equation must be performed on the opposite side to ensure that the equation is still in equilibrium.

$6X + 3 = 2X + 11$

To dispose of the 3 on the LHS, subtract 3 from the LHS and the RHS.

$6X + 3 - 3 = 2X + 11 - 3$

$6X = 2X + 8$

Subtract 2X from both sides of the equation:

$6X - 2X = 2X + 8 - 2X$

$4X = 8$

To solve X, the 4 must be removed from the 4X by dividing each side by 4.

$\dfrac{4X}{4} = \dfrac{8}{4}$

$X = 2$

11 Hierarchy of operations   The calculation procedure priority takes the following form from highest to lowest:

❏ Functions
❏ Calculations in parentheses ( )
❏ Powers and roots
❏ Multiplication and division
❏ Plus and minus

$$\dfrac{2 + 4(6^2 - 3 \times 2)\,(e^2 - 2 \times 2)^2}{\log 100}$$

$$= \frac{2 + 4(36-6)(7.39-4)^2}{2}$$
$$= \frac{2 + 4(30)(3.39)^2}{2}$$
$$= \frac{2 + 4(30)(11.49)}{2}$$
$$= \frac{2 + 4(344.7)}{2}$$
$$= \frac{2 + 1\,378.8}{2}$$
$$= \frac{1\,380.8}{2}$$
$$= 690.4$$

APPENDIX 3
Statistical and Interest Tables

CONTENTS

1 AREAS UNDER THE NORMAL CURVE 306
2 THE t-DISTRIBUTION 307
3 THE χ^2-DISTRIBUTION 308
4 THE F-DISTRIBUTION 309
5 THE STUDENTIZED RANGE 315
6 RANDOM NUMBERS . . . .
7 $(1+i)^n$ 316
8 $s_{\overline{n}|i}$ 321
9 $a_{\overline{n}|i}$ 325

TABLE 1:
The standard normal distribution

This table gives the area under the standard normal curve between 0 and z, i.e. $P[0 < Z < z]$

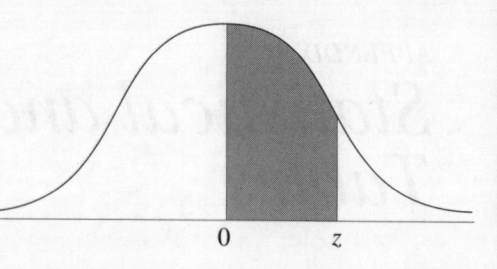

Z	0.00	0.01	0.02	0.03	0.04	0.05	0.06	0.07	0.08	0.09
0.0	0.0000	0.0040	0.0080	0.0120	0.0160	0.0199	0.0239	0.0279	0.0319	0.0359
0.1	0.0398	0.0438	0.0478	0.0517	0.0557	0.0596	0.0636	0.0675	0.0714	0.0753
0.2	0.0793	0.0832	0.0871	0.0910	0.0948	0.0987	0.1026	0.1064	0.1103	0.1141
0.3	0.1179	0.1217	0.1255	0.1293	0.1331	0.1368	0.1406	0.1443	0.1480	0.1517
0.4	0.1554	0.1591	0.1628	0.1664	0.1700	0.1736	0.1772	0.1808	0.1844	0.1879
0.5	0.1915	0.1950	0.1985	0.2019	0.2054	0.2088	0.2123	0.2157	0.2190	0.2224
0.6	0.2257	0.2291	0.2324	0.2357	0.2389	0.2422	0.2454	0.2486	0.2517	0.2549
0.7	0.2580	0.2611	0.2642	0.2673	0.2703	0.2734	0.2764	0.2793	0.2823	0.2852
0.8	0.2881	0.2910	0.2939	0.2967	0.2995	0.3023	0.3051	0.3078	0.3106	0.3133
0.9	0.3159	0.3186	0.3212	0.3238	0.3264	0.3289	0.3315	0.3340	0.3365	0.3389
1.0	0.3413	0.3438	0.3461	0.3485	0.3508	0.3531	0.3554	0.3557	0.3599	0.3621
1.1	0.3643	0.3665	0.3686	0.3708	0.3729	0.3749	0.3770	0.3790	0.3810	0.3830
1.2	0.3849	0.3869	0.3888	0.3907	0.3925	0.3944	0.3962	0.3980	0.3997	0.4015
1.3	0.4032	0.4049	0.4066	0.4082	0.4099	0.4115	0.4131	0.4147	0.4162	0.4177
1.4	0.4192	0.4207	0.4222	0.4236	0.4251	0.4265	0.4279	0.4292	0.4306	0.4319
1.5	0.4332	0.4345	0.4357	0.4370	0.4382	0.4394	0.4406	0.4418	0.4429	0.4441
1.6	0.4452	0.4463	0.4474	0.4484	0.4495	0.4505	0.4515	0.4525	0.4535	0.4545
1.7	0.4554	0.4564	0.4573	0.4582	0.4591	0.4599	0.4608	0.4616	0.4625	0.4633
1.8	0.4641	0.4649	0.4656	0.4664	0.4671	0.4678	0.4686	0.4693	0.4699	0.4706
1.9	0.4713	0.4719	0.4726	0.4732	0.4738	0.4744	0.4750	0.4756	0.4761	0.4767
2.0	0.4772	0.4778	0.4783	0.4788	0.4793	0.4798	0.4803	0.4808	0.4812	0.4817
2.1	0.4821	0.4826	0.4830	0.4834	0.4838	0.4842	0.4846	0.4850	0.4854	0.4857
2.2	0.4861	0.4864	0.4868	0.4871	0.4875	0.4878	0.4881	0.4884	0.4887	0.4890
2.3	0.48928	0.48956	0.48983	0.49010	0.49036	0.49061	0.49086	0.49111	0.49134	0.49158
2.4	0.49180	0.49202	0.49224	0.49245	0.49266	0.49286	0.49305	0.49324	0.49343	0.49361
2.5	0.49379	0.49396	0.49413	0.49430	0.49446	0.49461	0.49477	0.49492	0.49506	0.49520
2.6	0.49534	0.49547	0.49560	0.49573	0.49585	0.49598	0.49609	0.49621	0.49632	0.49643
2.7	0.49653	0.49664	0.49674	0.49683	0.49693	0.49702	0.49711	0.49720	0.49728	0.49736
2.8	0.49744	0.49752	0.49760	0.49767	0.49774	0.49781	0.49788	0.49795	0.49801	0.49807
2.9	0.49813	0.49819	0.49825	0.49831	0.49836	0.49841	0.49846	0.49851	0.49856	0.49861
3.0	0.49865	0.49869	0.49874	0.49878	0.49882	0.49886	0.49889	0.49893	0.49897	0.49900
3.1	0.49903	0.49906	0.49910	0.49913	0.49916	0.49918	0.49921	0.49924	0.49926	0.49929
3.2	0.49931	0.49934	0.49936	0.49938	0.49940	0.49942	0.49944	0.49946	0.49948	0.49950
3.3	0.49952	0.49953	0.49955	0.49957	0.49958	0.49960	0.49961	0.49962	0.49964	0.49965
3.4	0.49966	0.49968	0.49969	0.49970	0.49971	0.49972	0.49973	0.49974	0.49975	0.49976
3.5	0.49977	0.49978	0.49978	0.49979	0.49980	0.49981	0.49981	0.49982	0.49983	0.49983
3.6	0.49984	0.49985	0.49985	0.49986	0.49986	0.49987	0.49987	0.49988	0.49988	0.49989
3.7	0.49989	0.49990	0.49990	0.49990	0.49991	0.49991	0.49991	0.49992	0.49992	0.49992
3.8	0.49993	0.49993	0.49993	0.49994	0.49994	0.49994	0.49994	0.49995	0.49995	0.49995
3.9	0.49995	0.49995	0.49996	0.49996	0.49996	0.49996	0.49996	0.49996	0.49997	0.49997
4.0	0.49997	0.49997	0.49997	0.49997	0.49997	0.49997	0.49998	0.49998	0.49998	0.49998

TABLE 2:
The *t*-distribution
This table gives the value of *t* where *df* is the degrees of freedom

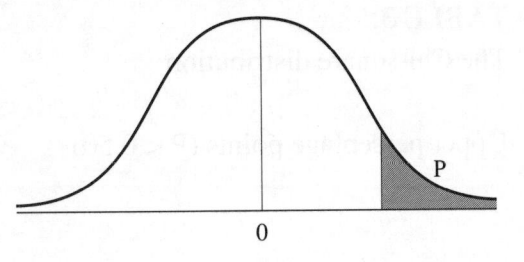

P	0.200	0.100	0.050	0.025	ONE-TAIL TEST 0.010	0.005	0.0025	0.0010	0.0005
					TWO-TAIL TEST				
	0.40	0.20	0.10	0.05	0.02	0.01	0.005	0.002	0.001
DF									
1	1.376	3.078	6.314	12.706	31.821	63.657	127.322	318.313	636.633
2	1.061	1.886	2.920	4.303	6.965	9.925	14.089	22.237	31.599
3	0.978	1.638	2.353	3.182	4.541	5.841	7.453	10.215	12.924
4	0.941	1.533	2.132	2.776	3.747	4.604	5.598	7.173	8.160
5	0.920	1.476	2.015	2.571	3.365	4.032	4.773	5.893	6.869
6	0.906	1.440	1.943	2.447	3.143	3.707	4.317	5.208	5.959
7	0.896	1.415	1.895	2.365	2.998	3.499	4.029	4.785	5.408
8	0.889	1.397	1.860	2.306	2.896	3.355	3.833	4.501	5.041
9	0.883	1.383	1.833	2.262	2.821	3.250	3.690	4.297	4.781
10	0.879	1.372	1.812	2.228	2.764	3.169	3.581	4.144	4.587
11	0.876	1.363	1.796	2.201	2.718	3.106	3.497	4.025	4.437
12	0.873	1.356	1.782	2.179	2.681	3.055	3.428	3.930	4.318
13	0.870	1.350	1.771	2.160	2.650	3.012	3.372	3.852	4.221
14	0.868	1.345	1.761	2.145	2.624	2.977	3.326	3.787	4.410
15	0.866	1.341	1.753	2.131	2.602	2.947	3.286	3.733	4.073
16	0.865	1.337	1.746	2.120	2.583	2.921	3.252	3.686	4.015
17	0.863	1.333	1.740	2.110	2.567	2.898	3.222	3.646	3.965
18	0.862	1.330	1.734	2.101	2.552	2.878	3.197	3.610	3.922
19	0.861	1.328	1.729	2.093	2.539	2.861	3.174	3.579	3.883
20	0.860	1.325	1.725	2.086	2.528	2.845	3.153	3.552	3.850
21	0.859	1.323	1.721	2.080	2.518	2.831	3.135	3.527	3.819
22	0.858	1.321	1.717	2.074	2.508	2.819	3.119	3.505	3.792
23	0.858	1.319	1.714	2.069	2.500	2.807	3.104	3.485	3.768
24	0.857	1.318	1.711	2.064	2.492	2.797	3.091	3.467	3.745
25	0.856	1.316	1.708	2.060	2.485	2.787	3.078	3.450	3.725
26	0.856	1.315	1.706	2.056	2.479	2.779	3.067	3.435	3.707
27	0.855	1.314	1.703	2.052	2.473	2.771	3.057	3.421	3.690
28	0.855	1.313	1.701	2.048	2.467	2.763	3.047	3.408	3.674
29	0.854	1.311	1.699	2.045	2.462	2.756	3.038	3.396	3.659
30	0.854	1.310	1.697	2.042	2.457	2.750	3.030	3.385	3.646
31	0.853	1.309	1.696	2.040	2.453	2.744	3.022	3.375	3.633
32	0.853	1.309	1.694	2.037	2.449	2.738	3.015	3.365	3.622
33	0.853	1.308	1.692	2.035	2.445	2.733	3.008	3.356	3.611
34	0.852	1.307	1.691	2.032	2.441	2.728	3.002	3.348	3.601
35	0.852	1.306	1.690	2.030	2.438	2.724	2.996	3.340	3.591
36	0.852	1.306	1.688	2.028	2.434	2.719	2.990	3.333	3.582
37	0.851	1.305	1.687	2.026	2.431	2.715	2.985	3.326	3.574
38	0.851	1.304	1.686	2.024	2.429	2.712	2.980	3.319	3.566
39	0.851	1.304	1.685	2.023	2.426	2.708	2.976	3.313	3.558
40	0.851	1.303	1.684	2.021	2.423	2.704	2.971	3.307	3.551
45	0.850	1.301	1.679	2.014	2.412	2.690	2.952	3.282	3.520
50	0.849	1.299	1.676	2.009	2.403	2.678	2.937	3.261	3.496
60	0.848	1.296	1.671	2.000	2.390	2.660	2.915	3.232	3.460
70	0.847	1.294	1.667	1.994	2.381	2.648	2.899	3.211	3.435
80	0.846	1.292	1.664	1.990	2.374	2.639	2.887	3.195	3.416
90	0.846	1.291	1.662	1.987	2.369	2.632	2.878	3.183	3.402
100	0.845	1.290	1.660	1.984	2.364	2.626	2.871	3.174	3.391
110	0.845	1.289	1.659	1.982	2.361	2.621	2.865	3.166	3.381
120	0.845	1.289	1.658	1.980	2.358	2.617	2.860	3.160	3.374
140	0.844	1.288	1.656	1.977	2.353	2.611	2.852	3.150	3.361
160	0.844	1.287	1.654	1.975	2.350	2.607	2.847	3.142	3.352
180	0.844	1.286	1.653	1.973	2.347	2.603	2.842	3.136	3.346
200	0.843	1.286	1.653	1.972	2.345	2.601	2.839	3.132	3.340
∞	0.841	1.282	1.645	1.960	2.327	2.576	2.807	3.091	3.291

TABLE 3:
The Chi-square distribution

Upper percentage points (P < 0.50)

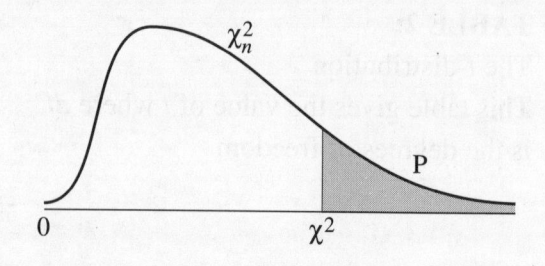

P\DF	0.200	0.100	0.050	0.025	0.01	0.005	0.0025	0.0010	0.0005
1	1.643	2.707	3.843	5.026	6.637	7.881	9.142	10.829	12.117
2	3.219	4.605	5.991	7.378	9.210	10.597	11.983	13.816	15.202
3	4.642	6.251	7.815	9.348	11.345	12.838	14.321	16.267	17.731
4	5.989	7.779	9.488	11.143	13.277	14.860	16.424	18.467	19.997
5	7.289	9.236	11.071	12.833	15.086	16.750	18.386	20.515	22.105
6	8.558	10.645	12.592	14.449	16.812	18.548	20.249	22.458	24.103
7	9.803	12.017	14.067	16.013	18.475	20.278	22.040	24.322	26.018
8	11.030	13.362	15.507	17.535	20.090	21.955	23.774	26.124	27.868
9	12.242	14.684	16.919	19.023	21.666	23.589	25.462	27.877	29.666
10	13.442	15.987	18.307	20.483	23.209	25.188	27.112	29.588	31.420
11	14.631	17.275	19.675	21.920	24.725	26.757	28.729	31.264	33.136
12	15.812	18.549	21.026	23.337	26.217	28.300	30.318	32.909	34.821
13	16.985	19.812	22.362	24.736	27.688	29.819	31.883	34.528	36.478
14	18.151	21.064	23.685	26.119	29.141	31.319	33.426	36.123	38.109
15	19.311	22.307	24.996	27.488	30.578	32.801	34.950	37.697	39.719
16	20.465	23.542	26.296	28.845	32.000	34.267	36.456	39.252	41.308
17	21.615	24.769	27.587	30.191	33.409	35.718	37.946	40.790	42.879
18	22.760	25.989	28.869	31.526	34.805	37.156	39.422	42.312	44.434
19	23.900	27.204	30.144	32.852	36.191	38.582	40.885	43.820	45.974
20	25.038	28.412	31.410	34.170	37.566	39.997	42.336	45.315	47.498
21	26.171	29.615	32.671	35.479	38.932	41.401	43.775	46.797	49.011
22	27.301	30.813	33.924	36.781	40.289	42.796	45.204	48.268	50.511
23	28.429	32.007	35.172	38.076	41.638	44.181	46.623	49.728	52.000
24	29.553	33.196	36.415	39.364	42.980	45.558	48.034	51.179	53.478
25	30.675	34.382	37.652	40.646	44.314	46.928	49.435	52.620	54.947
26	31.795	35.563	38.885	41.923	45.642	48.290	50.829	54.052	56.407
27	32.912	36.741	40.113	43.195	46.963	49.645	52.215	55.476	57.857
28	34.027	37.916	41.337	44.461	48.278	50.993	53.594	56.892	59.300
29	35.139	39.087	42.557	45.722	49.588	52.336	54.967	58.301	60.734
30	36.250	40.256	43.773	46.979	50.892	53.672	56.332	59.703	62.162
31	37.359	44.422	44.985	48.232	52.191	55.003	57.692	61.098	63.582
32	38.466	42.585	46.194	49.480	53.486	56.328	59.046	62.487	64.995
33	39.572	43.745	47.400	50.725	54.776	57.648	60.395	63.870	66.402
34	40.676	44.903	48.602	51.966	56.061	58.964	61.738	65.247	67.803
35	41.778	46.059	49.802	53.203	57.342	60.275	63.076	66.619	69.198
36	42.879	47.212	50.998	54.437	58.619	61.581	64.410	67.985	70.588
37	43.978	48.363	52.192	55.668	59.892	62.883	65.739	69.346	71.972
38	45.076	49.513	53.384	56.896	61.162	64.181	67.063	70.703	73.351
39	46.173	50.660	54.572	58.120	62.428	65.476	68.383	72.055	74.725
40	47.269	51.805	55.758	59.342	63.691	66.766	69.699	73.402	76.094
45	52.729	57.505	61.656	65.410	69.957	73.166	76.233	80.077	82.875
50	58.164	63.167	67.505	71.420	76.154	79.490	82.664	86.661	89.560
60	68.970	74.399	79.087	83.305	88.386	91.957	95.357	99.607	102.689
70	79.712	85.529	90.537	95.031	100.432	104.222	107.812	112.319	115.575
80	90.403	96.581	101.885	106.636	112.336	116.329	120.107	124.842	128.261
90	101.051	107.568	113.151	118.144	124.125	128.307	132.262	137.213	140.783
100	111.664	118.501	124.348	129.570	135.815	140.178	144.300	149.455	153.169
110	122.247	129.388	135.487	140.925	147.423	151.958	156.238	161.587	165.439
120	140.231	146.571	152.222	157.389	163.678	168.122	172.351	177.673	181.528
140	161.826	168.618	174.659	180.174	186.875	191.604	196.099	201.748	205.835
160	183.310	190.522	196.926	202.766	209.852	214.845	219.588	225.542	229.846
180	204.704	212.310	219.056	225.200	232.647	237.890	242.866	249.107	253.615

TABLE 4.1:

5 % Points of the F-distribution

This table gives values of $F^{0.05}_{NUM,DEN}$ where NUM = degrees of freedom for numerator, and DEN = degrees of freedom for denominator.

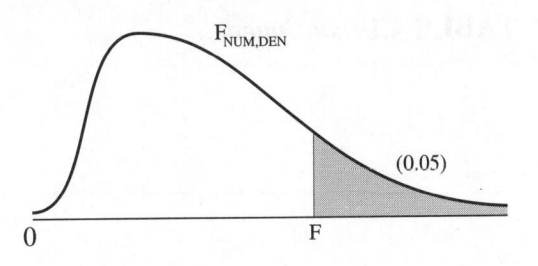

NUM\DEN	1	2	3	4	5	6	7	8	9	10	11	12	13	14
1	161	199	216	225	230	234	237	239	241	242	243	244	245	245
2	18.5	19.0	19.2	19.2	19.3	19.3	19.4	19.4	19.4	19.4	19.4	19.4	19.4	19.4
3	10.13	9.55	9.28	9.12	9.01	8.94	8.89	8.85	8.81	8.79	8.76	8.74	8.73	8.71
4	7.71	6.94	6.59	6.39	6.26	6.16	6.09	6.04	6.00	5.96	5.94	5.91	5.89	5.87
5	6.61	5.79	5.41	5.19	5.05	4.95	4.88	4.82	4.77	4.74	4.70	4.68	4.66	4.64
6	5.99	5.14	4.76	4.53	4.39	4.28	4.21	4.15	4.10	4.06	4.03	4.00	3.98	3.96
7	5.59	4.74	4.35	4.12	3.97	3.87	3.79	3.73	3.68	3.64	3.60	3.57	3.55	3.53
8	5.32	4.46	4.07	3.84	3.69	3.58	3.50	3.44	3.39	3.35	3.31	3.28	3.26	3.24
9	5.12	4.26	3.86	3.63	3.48	3.37	3.29	3.23	3.18	3.14	3.10	3.07	3.05	3.03
10	4.96	4.10	3.71	3.48	3.33	3.22	3.14	3.07	3.02	2.98	2.94	2.91	2.89	2.86
11	4.84	3.98	3.59	3.36	3.20	3.09	3.01	2.95	2.90	2.85	2.82	2.79	2.76	2.74
12	4.75	3.89	3.49	3.26	3.11	3.00	2.91	2.85	2.80	2.75	2.72	2.69	2.66	2.64
13	4.67	3.81	3.41	3.18	3.03	2.92	2.83	2.77	2.71	2.67	2.63	2.60	2.58	2.55
14	4.60	3.74	3.34	3.11	2.96	2.85	2.76	2.70	2.65	2.60	2.57	2.53	2.51	2.48
15	4.54	3.68	3.29	3.06	2.90	2.79	2.71	2.64	2.59	2.54	2.51	2.48	2.45	2.42
16	4.49	3.63	3.24	3.01	2.85	2.74	2.66	2.59	2.54	2.49	2.46	2.42	2.40	2.37
17	4.45	3.59	3.20	2.96	2.81	2.70	2.61	2.55	2.49	2.45	2.41	2.38	2.35	2.33
18	4.41	3.55	3.16	2.93	2.77	2.66	2.58	2.51	2.46	2.41	2.37	2.34	2.31	2.29
19	4.38	3.52	3.13	2.90	2.74	2.63	2.54	2.48	2.42	2.38	2.34	2.31	2.28	2.26
20	4.35	3.49	3.10	2.87	2.71	2.60	2.51	2.45	2.39	2.35	2.31	2.28	2.25	2.22
21	4.32	3.47	3.07	2.84	2.68	2.57	2.49	2.42	2.37	2.32	2.28	2.25	2.22	2.20
22	4.30	3.44	3.05	2.82	2.66	2.55	2.46	2.40	2.34	2.30	2.26	2.23	2.20	2.17
23	4.28	3.42	3.03	2.80	2.64	2.53	2.44	2.37	2.32	2.27	2.24	2.20	2.18	2.15
24	4.26	3.40	3.01	2.78	2.62	2.51	2.42	2.36	2.30	2.25	2.22	2.18	2.15	2.13
25	4.24	3.39	2.99	2.76	2.60	2.49	2.40	2.34	2.28	2.24	2.20	2.16	2.14	2.11
26	4.23	3.37	2.98	2.74	2.59	2.47	2.39	2.32	2.27	2.22	2.18	2.15	2.12	2.09
27	4.21	3.35	2.96	2.73	2.57	2.46	2.37	2.31	2.25	2.20	2.17	2.13	2.10	2.08
28	4.20	3.34	2.95	2.71	2.56	2.45	2.36	2.29	2.24	2.19	2.15	2.12	2.09	2.06
29	4.18	3.33	2.93	2.70	2.55	2.43	2.35	2.28	2.22	2.18	2.14	2.10	2.08	2.05
30	4.17	3.32	2.92	2.69	2.53	2.42	2.33	2.27	2.21	2.16	2.13	2.09	2.06	2.04
31	4.16	3.30	2.91	2.68	2.52	2.41	2.32	2.25	2.20	2.15	2.11	2.08	2.05	2.03
32	4.15	3.29	2.90	2.67	2.51	2.40	2.31	2.24	2.19	2.14	2.10	2.07	2.04	2.01
33	4.14	3.28	2.89	2.66	2.50	2.39	2.30	2.23	2.18	2.13	2.09	2.06	2.03	2.00
34	4.13	3.28	2.88	2.65	2.49	2.38	2.29	2.23	2.17	2.12	2.08	2.05	2.02	1.99
35	4.12	3.27	2.87	2.64	2.49	2.37	2.29	2.22	2.16	2.11	2.07	2.04	2.01	1.99
36	4.11	3.26	2.87	2.63	2.48	2.36	2.28	2.21	2.15	2.11	2.07	2.03	2.00	1.98
37	4.11	3.25	2.86	2.63	2.47	2.36	2.27	2.20	2.14	2.10	2.06	2.02	2.00	1.97
38	4.10	3.24	2.85	2.62	2.46	2.35	2.26	2.19	2.14	2.09	2.05	2.02	1.99	1.96
39	4.09	3.24	2.85	2.61	2.46	2.34	2.26	2.19	2.13	2.08	2.04	2.01	1.98	1.95
40	4.08	3.23	2.84	2.61	2.45	2.34	2.25	2.18	2.12	2.08	2.04	2.00	1.97	1.95
45	4.06	3.20	2.81	2.58	2.42	2.31	2.22	2.15	2.10	2.05	2.01	1.97	1.94	1.92
50	4.03	3.18	2.79	2.56	2.40	2.29	2.20	2.13	2.07	2.03	1.99	1.95	1.92	1.89
60	4.00	3.15	2.76	2.53	2.37	2.25	2.17	2.10	2.04	1.99	1.95	1.92	1.89	1.86
70	3.98	3.13	2.74	2.50	2.35	2.23	2.14	2.07	2.02	1.97	1.93	1.89	1.86	1.84
80	3.96	3.11	2.72	2.49	2.33	2.21	2.13	2.06	2.00	1.95	1.91	1.88	1.84	1.82
90	3.95	3.10	2.71	2.47	2.32	2.20	2.11	2.04	1.99	1.94	1.90	1.86	1.83	1.80
100	3.94	3.09	2.70	2.46	2.31	2.19	2.10	2.03	1.97	1.93	1.89	1.85	1.82	1.79
110	3.93	3.08	2.69	2.45	2.30	2.18	2.09	2.02	1.97	1.92	1.88	1.84	1.81	1.78
120	3.92	3.07	2.68	2.45	2.29	2.18	2.09	2.02	1.96	1.91	1.87	1.83	1.80	1.78
140	3.91	3.06	2.67	2.44	2.28	2.16	2.08	2.01	1.95	1.90	1.86	1.82	1.79	1.76
160	3.90	3.05	2.66	2.43	2.27	2.16	2.07	2.00	1.94	1.89	1.85	1.81	1.78	1.75
180	3.89	3.05	2.65	2.42	2.26	2.15	2.06	1.99	1.93	1.88	1.84	1.81	1.77	1.75
200	3.89	3.04	2.65	2.42	2.26	2.14	2.06	1.98	1.93	1.88	1.84	1.80	1.77	1.74
à	3.84	3.00	2.60	2.37	2.21	2.10	2.01	1.94	1.88	1.83	1.79	1.75	1.72	1.69

TABLE 4.1 (*continued*)

NUM\DEN	15	16	17	18	19	20	22	24	27	30	40	60	100	
1	246	246	247	247	248	248	249	249	250	250	251	252	253	256
2	19.4	19.4	19.4	19.4	19.4	19.4	19.5	19.5	19.5	19.5	19.5	19.5	19.5	19.5
3	8.70	8.69	8.68	8.67	8.67	8.66	8.65	8.64	8.63	8.62	8.59	8.57	8.55	8.53
4	5.86	5.84	5.83	5.82	5.81	5.80	5.79	5.77	5.76	5.75	5.72	5.69	5.66	5.63
5	4.62	4.60	4.59	4.58	4.57	4.56	4.54	4.53	4.51	4.50	4.46	4.43	4.41	4.36
6	3.94	3.92	3.91	3.90	3.88	3.87	3.86	3.84	3.82	3.81	3.77	3.74	3.71	3.67
7	3.51	3.49	3.48	3.47	3.46	3.44	3.43	3.41	3.39	3.38	3.34	3.30	3.27	3.23
8	3.22	3.20	3.19	3.17	3.16	3.15	3.13	3.12	3.10	3.08	3.04	3.01	2.97	2.93
9	3.01	2.99	2.97	2.96	2.95	2.94	2.92	2.90	2.88	2.86	2.83	2.79	2.76	2.71
10	2.85	2.83	2.81	2.80	2.79	2.77	2.75	2.74	2.72	2.70	2.66	2.62	2.59	2.54
11	2.72	2.70	2.69	2.67	2.66	2.65	2.63	2.61	2.59	2.57	2.53	2.49	2.46	2.40
12	2.62	2.60	2.58	2.57	2.56	2.54	2.52	2.51	2.48	2.47	2.43	2.38	2.35	2.30
13	2.53	2.51	2.50	2.48	2.47	2.46	2.44	2.42	2.40	2.38	2.34	2.30	2.26	2.21
14	2.46	2.44	2.43	2.41	2.40	2.39	2.37	2.35	2.33	2.31	2.27	2.22	2.19	2.13
15	2.40	2.38	2.37	2.35	2.34	2.33	2.31	2.29	2.27	2.25	2.20	2.16	2.12	2.07
16	2.35	2.33	2.32	2.30	2.29	2.28	2.25	2.24	2.21	2.19	2.15	2.11	2.07	2.01
17	2.31	2.29	2.27	2.26	2.24	2.23	2.21	2.19	2.17	2.15	2.10	2.06	2.02	1.96
18	2.27	2.25	2.23	2.22	2.20	2.19	2.17	2.15	2.13	2.11	2.06	2.02	1.98	1.92
19	2.23	2.21	2.20	2.18	2.17	2.16	2.13	2.11	2.09	2.07	2.03	1.98	1.94	1.88
20	2.20	2.18	2.17	2.15	2.14	2.12	2.10	2.08	2.06	2.04	1.99	1.95	1.91	1.84
21	2.18	2.16	2.14	2.12	2.11	2.10	2.07	2.05	2.03	2.01	1.96	1.92	1.88	1.81
22	2.15	2.13	2.11	2.10	2.08	2.07	2.05	2.03	2.00	1.98	1.94	1.89	1.85	1.78
23	2.13	2.11	2.09	2.08	2.06	2.05	2.02	2.01	1.98	1.96	1.91	1.86	1.82	1.76
24	2.11	2.09	2.07	2.05	2.04	2.03	2.00	1.98	1.96	1.94	1.89	1.84	1.80	1.73
25	2.09	2.07	2.05	2.04	2.02	2.01	1.98	1.96	1.94	1.92	1.87	1.82	1.78	1.71
26	2.07	2.05	2.03	2.02	2.00	1.99	1.97	1.95	1.92	1.90	1.85	1.80	1.76	1.69
27	2.06	2.04	2.02	2.00	1.99	1.97	1.95	1.93	1.90	1.88	1.84	1.79	1.74	1.67
28	2.04	2.02	2.00	1.99	1.97	1.96	1.93	1.91	1.89	1.87	1.82	1.77	1.73	1.65
29	2.03	2.01	1.99	1.97	1.96	1.94	1.92	1.90	1.88	1.85	1.81	1.75	1.71	1.64
30	2.01	1.99	1.98	1.96	1.95	1.93	1.91	1.89	1.86	1.84	1.79	1.74	1.70	1.62
31	2.00	1.98	1.96	1.95	1.93	1.92	1.90	1.88	1.85	1.83	1.78	1.73	1.68	1.61
32	1.99	1.97	1.95	1.94	1.92	1.91	1.88	1.86	1.84	1.82	1.77	1.71	1.67	1.59
33	1.98	1.96	1.94	1.93	1.91	1.90	1.87	1.85	1.83	1.81	1.76	1.70	1.66	1.58
34	1.97	1.95	1.93	1.92	1.90	1.89	1.86	1.84	1.82	1.80	1.75	1.69	1.65	1.57
35	1.96	1.94	1.92	1.91	1.89	1.88	1.85	1.83	1.81	1.79	1.74	1.68	1.63	1.56
36	1.95	1.93	1.92	1.90	1.88	1.87	1.85	1.82	1.80	1.78	1.73	1.67	1.62	1.55
37	1.95	1.93	1.91	1.89	1.88	1.86	1.84	1.82	1.79	1.77	1.72	1.66	1.62	1.54
38	1.94	1.92	1.90	1.88	1.87	1.85	1.83	1.81	1.78	1.76	1.71	1.65	1.61	1.53
39	1.93	1.91	1.89	1.88	1.86	1.85	1.82	1.80	1.77	1.75	1.70	1.65	1.60	1.52
40	1.92	1.90	1.89	1.87	1.85	1.84	1.81	1.79	1.77	1.74	1.69	1.64	1.59	1.51
45	1.89	1.87	1.86	1.84	1.82	1.81	1.78	1.76	1.73	1.71	1.66	1.60	1.55	1.47
50	1.87	1.85	1.83	1.81	1.80	1.78	1.76	1.74	1.71	1.69	1.63	1.58	1.53	1.44
60	1.84	1.82	1.80	1.78	1.76	1.75	1.72	1.70	1.67	1.65	1.59	1.53	1.48	1.39
70	1.81	1.79	1.77	1.75	1.74	1.72	1.70	1.67	1.65	1.62	1.57	1.50	1.45	1.35
80	1.79	1.77	1.75	1.73	1.72	1.70	1.68	1.65	1.63	1.60	1.54	1.48	1.43	1.32
90	1.78	1.76	1.74	1.72	1.70	1.69	1.66	1.64	1.61	1.59	1.53	1.46	1.41	1.30
100	1.77	1.75	1.73	1.71	1.69	1.68	1.65	1.63	1.60	1.57	1.52	1.45	1.39	1.28
110	1.76	1.74	1.72	1.70	1.68	1.67	1.64	1.62	1.59	1.56	1.50	1.44	1.37	1.27
120	1.75	1.73	1.71	1.69	1.67	1.66	1.63	1.61	1.58	1.55	1.50	1.43	1.36	1.25
140	1.74	1.72	1.70	1.68	1.66	1.65	1.62	1.60	1.57	1.54	1.48	1.41	1.35	1.23
160	1.73	1.71	1.69	1.67	1.65	1.64	1.61	1.59	1.56	1.53	1.47	1.40	1.34	1.21
180	1.72	1.70	1.68	1.66	1.64	1.63	1.60	1.58	1.55	1.52	1.46	1.39	1.33	1.20
200	1.72	1.69	1.67	1.66	1.64	1.62	1.60	1.57	1.54	1.52	1.46	1.39	1.32	1.19
à	1.67	1.64	1.62	1.60	1.59	1.57	1.54	1.52	1.49	1.46	1.39	1.32	1.24	1.00

TABLE 4.2:
2½ % points of the F-distribution

This table gives values of $F^{0.25}_{NUM,DEN}$ where NUM = degrees of freedom for numerator, and DEN = degrees of freedom for denominator.

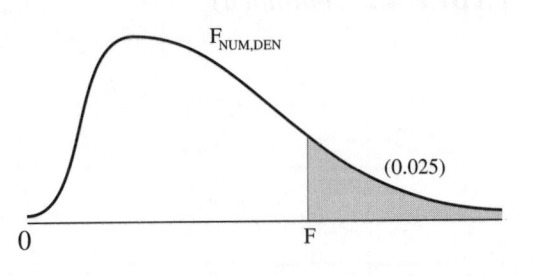

NUM\DEN	1	2	3	4	5	6	7	8	9	10	11	12	13	14
1	648	800	864	900	922	937	948	957	963	969	973	977	980	983
2	38.5	39.0	39.2	39.2	39.3	39.3	39.4	39.4	39.4	39.4	39.4	39.4	39.4	39.4.
3	17.44	16.04	15.44	15.10	14.88	14.73	14.62	14.54	14.47	14.42	14.37	14.34	14.30	14.28
4	12.22	10.65	9.98	9.60	9.36	9.20	9.07	8.98	8.90	8.84	8.79	8.75	8.71	8.68
5	10.01	8.43	7.76	7.39	7.15	6.98	6.85	6.76	6.68	6.62	6.57	6.52	6.49	6.46
6	8.81	7.26	6.60	6.23	5.99	5.82	5.70	5.60	5.52	5.46	5.41	5.37	5.33	5.30
7	8.07	6.54	5.89	5.52	5.29	5.12	4.99	4.90	4.82	4.76	4.71	4.67	4.63	4.60
8	7.57	6.06	5.42	5.05	4.82	4.65	4.53	4.43	4.36	4.30	4.24	4.20	4.16	4.13
9	7.21	5.71	5.08	4.72	4.48	4.32	4.20	4.10	4.03	3.96	3.91	3.87	3.83	3.80
10	6.94	5.46	4.83	4.47	4.24	4.07	3.95	3.85	3.78	3.72	3.66	3.62	3.58	3.55
11	6.72	5.26	4.63	4.28	4.04	3.88	3.76	3.66	3.59	3.53	3.47	3.43	3.39	3.36
12	6.55	5.10	4.47	4.12	3.89	3.73	3.61	3.51	3.44	3.37	3.32	3.28	3.24	3.21
13	6.41	4.97	4.35	4.00	3.77	3.60	3.48	3.39	3.31	3.25	3.20	3.15	3.12	3.08
14	6.30	4.86	4.24	3.89	3.66	3.50	3.38	3.29	3.21	3.15	3.09	3.05	3.01	2.98
15	6.20	4.77	4.15	3.80	3.58	3.41	3.29	3.20	3.12	3.06	3.01	2.96	2.92	2.89
16	6.12	4.69	4.08	3.73	3.50	3.34	3.22	3.12	3.05	2.99	2.93	2.89	2.85	2.82
17	6.04	4.62	4.01	3.66	3.44	3.28	3.16	3.06	2.98	2.92	2.87	2.82	2.79	2.75
18	5.98	4.56	3.95	3.61	3.38	3.22	3.10	3.01	2.93	2.87	2.81	2.77	2.73	2.70
19	5.92	4.51	3.90	3.56	3.33	3.17	3.05	2.96	2.88	2.82	2.76	2.72	2.68	2.65
20	5.87	4.46	3.86	3.51	3.29	3.13	3.01	2.91	2.84	2.77	2.72	2.68	2.64	2.60
21	5.83	4.42	3.82	3.48	3.25	3.09	2.97	2.87	2.80	2.73	2.68	2.64	2.60	2.56
22	5.79	4.38	3.78	3.44	3.22	3.05	2.93	2.84	2.76	2.70	2.65	2.60	2.56	2.53
23	5.75	4.35	3.75	3.41	3.18	3.02	2.90	2.81	2.73	2.67	2.62	2.57	2.53	2.50
24	5.72	4.32	3.72	3.38	3.15	2.99	2.87	2.78	2.70	2.64	2.59	2.54	2.50	2.47
25	5.69	4.29	3.69	3.35	3.13	2.97	2.85	2.75	2.68	2.61	2.56	2.51	2.48	2.44
26	5.66	4.27	3.67	3.33	3.10	2.94	2.82	2.73	2.65	2.59	2.54	2.49	2.45	2.42
27	5.63	4.24	3.65	3.31	3.08	2.92	2.80	2.71	2.63	2.57	2.51	2.47	2.43	2.39
28	5.61	4.22	3.63	3.29	3.06	2.90	2.78	2.69	2.61	2.55	2.49	2.45	2.41	2.37
29	5.59	4.20	3.61	3.27	3.04	2.88	2.76	2.67	2.59	2.53	2.48	2.43	2.39	2.36
30	5.57	4.18	3.59	3.25	3.03	2.87	2.75	2.65	2.57	2.51	2.46	2.41	2.37	2.34
31	5.55	4.16	3.57	3.23	3.01	2.85	2.73	2.64	2.56	2.50	2.44	2.40	2.36	2.32
32	5.53	4.15	3.56	3.22	3.00	2.84	2.71	2.62	2.54	2.48	2.43	2.38	2.34	2.31
33	5.51	4.13	3.54	3.20	2.98	2.82	2.70	2.61	2.53	2.47	2.41	2.37	2.33	2.29
34	5.50	4.12	3.53	3.19	2.97	2.81	2.69	2.59	2.52	2.45	2.40	2.35	2.31	2.28
35	5.48	4.11	3.52	3.18	2.96	2.80	2.68	2.58	2.50	2.44	2.39	2.34	2.30	2.27
36	5.47	4.09	3.50	3.17	2.94	2.78	2.66	2.57	2.49	2.43	2.37	2.33	2.29	2.25
37	5.46	4.08	3.49	3.16	2.93	2.77	2.65	2.56	2.48	2.42	2.36	2.32	2.28	2.24
38	5.45	4.07	3.48	3.15	2.92	2.76	2.64	2.55	2.47	2.41	2.35	2.31	2.27	2.23
39	5.43	4.06	3.47	3.14	2.91	2.75	2.63	2.54	2.46	2.40	2.34	2.30	2.26	2.22
40	5.42	4.05	3.46	3.13	2.90	2.74	2.62	2.53	2.45	2.39	2.33	2.29	2.25	2.21
45	5.38	4.01	3.42	3.09	2.86	2.70	2.58	2.49	2.41	2.35	2.29	2.25	2.21	2.17
50	5.34	3.97	3.39	3.05	2.83	2.67	2.55	2.46	2.38	2.32	2.26	2.22	2.18	2.14
60	5.29	3.93	3.34	3.01	2.79	2.63	2.51	2.41	2.33	2.27	2.22	2.17	2.13	2.09
70	5.25	3.89	3.31	2.97	2.75	2.59	2.47	2.38	2.30	2.24	2.18	2.14	2.10	2.06
80	5.22	3.86	3.28	2.95	2.73	2.57	2.45	2.35	2.28	2.21	2.16	2.11	2.07	2.03
90	5.20	3.84	3.26	2.93	2.71	2.55	2.43	2.34	2.26	2.19	2.14	2.09	2.05	2.02
100	5.18	3.83	3.25	2.92	2.70	2.54	2.42	2.32	2.24	2.18	2.12	2.08	2.04	2.00
110	5.16	3.82	3.24	2.90	2.68	2.53	2.40	2.31	2.23	2.17	2.11	2.07	2.02	1.99
120	5.15	3.80	3.23	2.89	2.67	2.52	2.39	2.30	2.22	2.16	2.10	2.05	2.01	1.98
140	5.13	3.79	3.21	2.88	2.66	2.50	2.38	2.28	2.21	2.14	2.09	2.04	2.00	1.96
160	5.12	3.78	3.20	2.87	2.65	2.49	2.37	2.27	2.19	2.13	2.07	2.03	1.99	1.95
180	5.11	3.77	3.19	2.86	2.64	2.48	2.36	2.26	2.19	2.12	2.07	2.02	1.98	1.94
200	5.10	3.76	3.18	2.85	2.63	2.47	2.35	2.26	2.18	2.11	2.06	2.01	1.97	1.93
à	5.03	3.69	3.12	2.79	2.57	2.41	2.29	2.19	2.11	2.05	1.99	1.94	1.90	1.87

TABLE 4.2 (*continued*)

NUM\DEN	15	16	17	18	19	20	22	24	27	30	40	60	100	
1	985	987	989	990	992	993	995	997	1000	1001	1006	1010	1013	1024
2	39.4	39.4	39.4	39.4	39.4	39.4	39.5	39.5	39.5	39.5	39.5	39.5	39.5	39.5
3	14.25	14.23	14.21	14.20	14.18	14.17	14.14	14.12	14.10	14.08	14.04	13.99	13.96	13.90
4	8.66	8.63	8.61	8.59	8.58	8.56	8.53	8.51	8.48	8.46	8.41	8.36	8.32	8.26
5	6.43	6.40	6.38	6.36	6.34	6.33	6.30	6.28	6.25	6.23	6.18	6.12	6.08	6.02
6	5.27	5.24	5.22	5.20	5.18	5.17	5.14	5.12	5.09	5.07	5.01	4.96	4.92	4.85
7	4.57	4.54	4.52	4.50	4.48	4.47	4.44	4.41	4.39	4.36	4.31	4.25	4.21	4.14
8	4.10	4.08	4.05	4.03	4.02	4.00	3.97	3.95	3.92	3.89	3.84	3.78	3.74	3.67
9	3.77	3.74	3.72	3.70	3.68	3.67	3.64	3.61	3.58	3.56	3.51	3.45	3.40	3.33
10	3.52	3.50	3.47	3.45	3.44	3.42	3.39	3.37	3.34	3.31	3.26	3.20	3.15	3.08
11	3.33	3.30	3.28	3.26	3.24	3.23	3.20	3.17	3.14	3.12	3.06	3.00	2.96	2.88
12	3.18	3.15	3.13	3.11	3.09	3.07	3.04	3.02	2.99	2.96	2.91	2.85	2.80	2.72
13	3.05	3.03	3.00	2.98	2.96	2.95	2.92	2.89	2.86	2.84	2.78	2.72	2.67	2.60
14	2.95	2.92	2.90	2.88	2.86	2.84	2.81	2.79	2.76	2.73	2.67	2.61	2.56	2.49
15	2.86	2.84	2.81	2.79	2.77	2.76	2.73	2.70	2.67	2.64	2.59	2.52	2.47	2.40
16	2.79	2.76	2.74	2.72	2.70	2.68	2.65	2.63	2.59	2.57	2.51	2.45	2.40	2.32
17	2.72	2.70	2.67	2.65	2.63	2.62	2.59	2.56	2.53	2.50	2.44	2.38	2.33	2.25
18	2.67	2.64	2.62	2.60	2.58	2.56	2.53	2.50	2.47	2.44	2.38	2.32	2.27	2.19
19	2.62	2.59	2.57	2.55	2.53	2.51	2.48	2.45	2.42	2.39	2.33	2.27	2.22	2.13
20	2.57	2.55	2.52	2.50	2.48	2.46	2.43	2.41	2.38	2.35	2.29	2.22	2.17	2.09
21	2.53	2.51	2.48	2.46	2.44	2.42	2.39	2.37	2.33	2.31	2.25	2.18	2.13	2.04
22	2.50	2.47	2.45	2.43	2.41	2.39	2.36	2.33	2.30	2.27	2.21	2.14	2.09	2.00
23	2.47	2.44	2.42	2.39	2.37	2.36	2.33	2.30	2.27	2.24	2.18	2.11	2.06	1.97
24	2.44	2.41	2.39	2.36	2.35	2.33	2.30	2.27	2.24	2.21	2.15	2.08	2.02	1.94
25	2.41	2.28	2.36	2.34	2.32	2.30	2.27	2.24	2.21	2.18	2.12	2.05	2.00	1.91
26	2.39	2.36	2.34	2.31	2.29	2.28	2.24	2.22	2.18	2.16	2.09	2.03	1.97	1.88
27	2.36	2.34	2.31	2.29	2.27	2.25	2.22	2.19	2.16	2.13	2.07	2.00	1.94	1.85
28	2.34	2.32	2.29	2.27	2.25	2.23	2.20	2.17	2.14	2.11	2.05	1.98	1.92	1.83
29	2.32	2.30	2.27	2.25	2.23	2.21	2.18	2.15	2.12	2.09	2.03	1.96	1.90	1.81
30	2.31	2.28	2.26	2.23	2.21	2.20	2.16	2.14	2.10	2.07	2.01	1.94	1.88	1.79
31	2.29	2.26	2.24	2.22	2.20	2.18	2.15	2.12	2.08	2.06	1.99	1.92	1.86	1.77
32	2.28	2.25	2.22	2.20	2.18	2.16	2.13	2.10	2.07	2.04	1.98	1.91	1.85	1.75
33	2.26	2.23	2.21	2.19	2.17	2.15	2.12	2.09	2.05	2.03	1.96	1.89	1.83	1.73
34	2.25	2.22	2.20	2.17	2.15	2.13	2.10	2.07	2.04	2.01	1.95	1.88	1.82	1.72
35	2.23	2.21	2.18	2.16	2.14	2.12	2.09	2.06	2.03	2.00	1.93	1.86	1.80	1.70
36	2.22	2.20	2.17	2.15	2.13	2.11	2.08	2.05	2.01	1.99	1.92	1.85	1.79	1.69
37	2.21	2.18	2.16	2.14	2.12	2.10	2.07	2.04	2.00	1.97	1.91	1.84	1.77	1.67
38	2.20	2.17	2.15	2.13	2.11	2.09	2.05	2.03	1.99	1.96	1.90	1.82	1.76	1.66
39	2.19	2.16	2.14	2.12	2.10	2.08	2.04	2.02	1.98	1.95	1.89	1.81	1.75	1.65
40	2.18	2.15	2.13	2.11	2.09	2.07	2.03	2.01	1.97	1.94	1.88	1.80	1.74	1.64
45	2.14	2.11	2.09	2.07	2.04	2.03	1.99	1.96	1.93	1.90	1.83	1.76	1.69	1.59
50	2.11	2.08	2.06	2.03	2.01	1.99	1.96	1.93	1.90	1.87	1.80	1.72	1.66	1.55
60	2.06	2.03	2.01	1.98	1.96	1.94	1.91	1.88	1.85	1.82	1.74	1.67	1.62	1.48
70	2.03	2.00	1.97	1.95	1.93	1.91	1.88	1.85	1.81	1.78	1.71	1.63	1.58	1.44
80	2.00	1.97	1.95	1.92	1.90	1.88	1.85	1.82	1.78	1.75	1.68	1.60	1.54	1.40
90	1.98	1.95	1.93	1.91	1.88	1.86	1.83	1.80	1.76	1.73	1.66	1.58	1.52	1.37
100	1.97	1.94	1.91	1.89	1.87	1.85	1.81	1.78	1.75	1.71	1.64	1.56	1.50	1.35
110	1.96	1.93	1.90	1.88	1.86	1.84	1.80	1.77	1.73	1.70	1.63	1.54	1.48	1.33
120	1.94	1.92	1.89	1.87	1.84	1.82	1.79	1.76	1.72	1.69	1.61	1.53	1.47	1.31
140	1.93	1.90	1.87	1.85	1.83	1.81	1.77	1.74	1.70	1.67	1.60	1.51	1.44	1.28
160	1.92	1.89	1.86	1.84	1.82	1.80	1.76	1.73	1.69	1.66	1.58	1.50	1.43	1.26
180	1.91	1.88	1.85	1.83	1.81	1.79	1.75	1.72	1.68	1.65	1.57	1.48	1.41	1.24
200	1.90	1.87	1.84	1.82	1.80	1.78	1.74	1.71	1.67	1.64	1.56	1.47	1.40	1.23
à	1.83	1.80	1.78	1.75	1.73	1.71	1.67	1.64	1.60	1.57	1.48	1.39	1.30	1.00

TABLE 4.3:
1 % Points of the F-distribution

This table gives values of $F^{0.01}_{NUM,DEN}$ where NUM = degrees of freedom for numerator, and DEN = degrees of freedom for denominator.

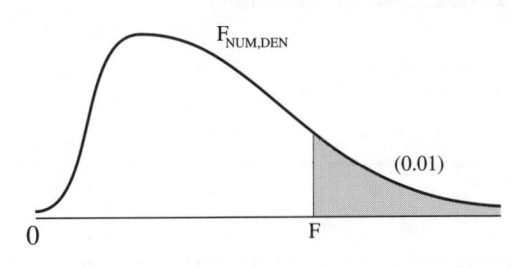

NUM\DEN	1	2	3	4	5	7	8	9	10	11	12	13	14
1	4052	4999	5403	5625	5764	5859	5928	5981	6022	6056	6083	6106	6126	6143
2	98.5	99.0	99.2	99.2	99.3	99.3	99.4	99.4	99.4	99.4	99.4	99.4	99.4	99.4
3	34.12	30.82	29.46	28.71	28.24	27.91	27.67	27.49	27.35	27.23	27.13	27.05	26.98	26.92
4	21.20	18.00	16.69	15.98	15.52	15.21	14.98	14.80	14.66	14.55	14.45	14.37	14.31	14.25
5	16.26	13.27	12.06	11.39	10.97	10.67	10.46	10.29	10.16	10.05	9.96	9.89	9.82	9.77
6	13.75	10.92	9.78	9.15	8.75	8.47	8.26	8.10	7.98	7.87	7.79	7.72	7.66	7.60
7	12.25	9.55	8.45	7.85	7.46	7.19	6.99	6.84	6.72	6.62	6.54	6.47	6.41	6.36
8	11.26	8.65	7.59	7.01	6.63	6.37	6.18	6.03	5.91	5.81	5.73	5.67	5.61	5.56
9	10.56	8.02	6.99	6.42	6.06	5.80	5.61	5.47	5.35	5.26	5.18	5.11	5.05	5.01
10	10.04	7.56	6.55	5.99	5.64	5.39	5.20	5.06	4.94	4.85	4.77	4.71	4.65	4.60
11	9.65	7.21	6.22	5.67	5.32	5.07	4.89	4.74	4.63	4.54	4.46	4.40	4.34	4.29
12	9.33	6.93	5.95	5.41	5.06	4.82	4.64	4.50	4.39	4.30	4.22	4.16	4.10	4.05
13	9.07	6.70	5.74	5.21	4.86	4.62	4.44	4.30	4.19	4.10	4.02	3.96	3.91	3.86
14	8.86	6.51	5.56	5.04	4.69	4.46	4.28	4.14	4.03	3.94	3.86	3.80	3.75	3.70
15	8.68	6.36	5.42	4.89	4.56	4.32	4.14	4.00	3.89	3.80	3.73	3.67	3.61	3.56
16	8.53	6.23	5.29	4.77	4.44	4.20	4.03	3.89	3.78	3.69	3.62	3.55	3.50	3.45
17	8.40	6.11	5.19	4.67	4.34	4.10	3.93	3.79	3.68	3.59	3.52	3.46	3.40	3.35
18	8.29	6.01	5.09	4.58	4.25	4.01	3.84	3.71	3.60	3.51	3.43	3.37	3.32	3.27
19	8.18	5.93	5.01	4.50	4.17	3.94	3.77	3.63	3.52	3.43	3.36	3.30	3.24	3.19
20	8.10	5.85	4.94	4.43	4.10	3.87	3.70	3.56	3.46	3.37	3.29	3.23	3.18	3.13
21	8.02	5.78	4.87	4.37	4.04	3.81	3.64	3.51	3.40	3.31	3.24	3.17	3.12	3.07
22	7.95	5.72	4.82	4.31	3.99	3.76	3.59	3.45	3.35	3.26	3.18	3.12	3.07	3.02
23	7.88	5.66	4.76	4.26	3.94	3.71	3.54	3.41	3.30	3.21	3.14	3.07	3.02	2.97
24	7.82	5.61	4.72	4.22	3.90	3.67	3.50	3.36	3.26	3.17	3.09	3.03	2.98	2.93
25	7.77	5.57	4.68	4.18	3.85	3.63	3.46	3.32	3.22	3.13	3.06	2.99	2.94	2.89
26	7.72	5.53	4.64	4.14	3.82	3.59	3.42	3.29	3.18	3.09	3.02	2.96	2.90	2.86
27	7.68	5.49	4.60	4.11	3.78	3.56	3.39	3.26	3.15	3.06	2.99	2.93	2.87	2.82
28	7.64	5.45	4.57	4.07	3.75	3.53	3.36	3.23	3.12	3.03	2.96	2.90	2.84	2.79
29	7.60	5.42	4.54	4.04	3.73	3.50	3.33	3.20	3.09	3.00	2.93	2.87	2.81	2.77
30	7.56	5.39	4.51	4.02	3.70	3.47	3.30	3.17	3.07	2.98	2.91	2.84	2.79	2.74
31	7.53	5.36	4.48	3.99	3.67	3.45	3.28	3.15	3.04	2.96	2.88	2.82	2.77	2.72
32	7.50	5.34	4.46	3.97	3.65	3.43	3.26	3.13	3.02	2.93	2.86	2.80	2.74	2.70
33	7.47	5.31	4.44	3.95	3.63	3.41	3.24	3.11	3.00	2.91	2.84	2.78	2.72	2.68
34	7.44	5.29	4.42	3.93	3.61	3.39	3.22	3.09	2.98	2.89	2.82	2.76	2.70	2.66
35	7.42	5.27	4.40	3.91	3.59	3.37	3.20	3.07	2.96	2.88	2.80	2.74	2.69	2.64
36	7.40	5.25	4.38	3.89	3.57	3.35	3.18	3.05	2.95	2.86	2.79	2.72	2.67	2.62
37	7.37	5.23	4.36	3.87	3.56	3.33	3.17	3.04	2.93	2.84	2.77	2.71	2.65	2.61
38	7.35	5.21	4.34	3.86	3.54	3.32	3.15	3.02	2.92	2.83	2.75	2.69	2.64	2.59
39	7.33	5.19	4.33	3.84	3.53	3.30	3.14	3.01	2.90	2.81	2.74	2.68	2.62	2.58
40	7.31	5.18	4.31	3.83	3.51	3.29	3.12	2.99	2.89	2.80	2.73	2.66	2.61	2.56
45	7.23	5.11	4.25	3.77	3.45	3.23	3.07	2.94	2.83	2.74	2.67	2.61	2.55	2.51
50	7.17	5.06	4.20	3.72	3.41	3.19	3.02	2.89	2.78	2.70	2.63	2.56	2.51	2.46
60	7.08	4.98	4.13	3.65	3.34	3.12	2.95	2.82	2.72	2.63	2.56	2.50	2.44	2.39
70	7.01	4.92	4.07	3.60	3.29	3.07	2.91	2.78	2.67	2.59	2.51	2.45	2.40	2.35
80	6.96	4.88	4.04	3.56	3.26	3.04	2.87	2.74	2.64	2.55	2.48	2.42	2.36	2.31
90	6.93	4.85	4.01	3.53	3.23	3.01	2.84	2.72	2.61	2.52	2.45	2.39	2.33	2.29
100	6.90	4.82	3.98	3.51	3.21	2.99	2.82	2.69	2.59	2.50	2.43	2.37	2.31	2.27
110	6.87	4.80	3.96	3.49	3.19	2.97	2.81	2.68	2.57	2.49	2.41	2.35	2.30	2.25
120	6.85	4.79	3.95	3.48	3.17	2.96	2.79	2.66	2.56	2.47	2.40	2.34	2.28	2.23
140	6.82	4.76	3.92	3.46	3.15	2.93	2.77	2.64	2.54	2.45	2.38	2.31	2.26	2.21
160	6.80	4.74	3.91	3.44	3.13	2.92	2.75	2.62	2.52	2.43	2.36	2.30	2.24	2.20
180	6.78	4.73	3.89	3.43	3.12	2.90	2.74	2.61	2.51	2.42	2.35	2.28	2.23	2.18
200	6.76	4.71	3.88	3.41	3.11	2.89	2.73	2.60	2.50	2.41	2.34	2.27	2.22	2.17
à	6.64	4.61	3.78	3.32	3.02	2.80	2.64	2.51	2.41	2.32	2.25	2.18	2.13	2.08

TABLE 4.3 (*continued*)

NUM\DEN	15	16	17	18	19	20	22	24	27	30	40	60	100
1	6157	6170	6181	6192	6201	6209	6223	6235	6249	6261	6287	6313	6334	6403
2	99.4	99.4	99.4	99.4	99.4	99.4	99.4	99.5	99.5	99.5	99.5	99.5	99.5	99.5
3	26.87	26.83	26.79	26.75	26.72	26.69	26.64	26.60	26.55	26.50	26.41	26.32	26.24	26.13
4	14.20	14.15	14.11	14.08	14.05	14.02	13.97	13.93	13.88	13.84	13.75	13.65	13.58	13.46
5	9.72	9.68	9.64	9.61	9.58	9.55	9.51	9.47	9.42	9.38	9.29	9.20	9.13	9.02
6	7.56	7.52	7.48	7.45	7.42	7.40	7.35	7.31	7.27	7.23	7.14	7.06	6.99	6.88
7	6.31	6.28	6.24	6.21	6.18	6.16	6.11	6.07	6.03	5.99	5.91	5.82	5.75	5.65
8	5.52	5.48	5.44	5.41	5.38	5.36	5.32	5.28	5.23	5.20	5.12	5.03	4.96	4.86
9	4.96	4.92	4.89	4.86	4.83	4.81	4.77	4.73	4.68	4.65	4.57	4.48	4.41	4.31
10	4.56	4.52	4.49	4.46	4.43	4.41	4.36	4.33	4.28	4.25	4.17	4.08	4.01	3.91
11	4.25	4.21	4.18	4.15	4.12	4.10	4.06	4.02	3.98	3.94	3.86	3.78	3.71	3.60
12	4.01	3.97	3.94	3.91	3.88	3.86	3.82	3.78	3.74	3.70	3.62	3.54	3.47	3.36
13	3.82	3.78	3.75	3.72	3.69	3.66	3.62	3.59	3.54	3.51	3.43	3.34	3.27	3.17
14	3.66	3.62	3.59	3.56	3.53	3.51	3.46	3.43	3.38	3.35	3.27	3.18	3.11	3.00
15	3.52	3.49	3.45	3.42	3.40	3.37	3.33	3.29	3.25	3.21	3.13	3.05	2.98	2.87
16	3.41	3.37	3.34	3.31	3.28	3.26	3.22	3.18	3.14	3.10	3.02	2.93	2.86	2.75
17	3.31	3.27	3.24	3.21	3.19	3.16	3.12	3.08	3.04	3.00	2.92	2.83	2.76	2.65
18	3.23	3.19	3.16	3.13	3.10	3.08	3.03	3.00	2.95	2.92	2.84	2.75	2.68	2.57
19	3.15	3.12	3.08	3.05	3.03	3.00	2.96	2.92	2.88	2.84	2.76	2.67	2.60	2.49
20	3.09	3.05	3.02	2.99	2.96	2.94	2.90	2.86	2.81	2.78	2.69	2.61	2.54	2.42
21	3.03	2.99	2.96	2.93	2.90	2.88	2.84	2.80	2.76	2.72	2.64	2.55	2.48	2.36
22	2.98	2.94	2.91	2.88	2.85	2.83	2.78	2.75	2.70	2.67	2.58	2.50	2.42	2.31
23	2.93	2.89	2.86	2.83	2.80	2.78	2.74	2.70	2.66	2.62	2.54	2.45	2.37	2.26
24	2.89	2.85	2.82	2.79	2.76	2.74	2.70	2.66	2.61	2.58	2.49	2.40	2.33	2.21
25	2.85	2.81	2.78	2.75	2.72	2.70	2.66	2.62	2.58	2.54	2.45	2.36	2.29	2.17
26	2.81	2.78	2.75	2.72	2.69	2.66	2.62	2.58	2.54	2.50	2.42	2.33	2.25	2.13
27	2.78	2.75	2.71	2.68	2.66	2.63	2.59	2.55	2.51	2.47	2.38	2.29	2.22	2.10
28	2.75	2.72	2.68	2.65	2.63	2.60	2.56	2.52	2.48	2.44	2.35	2.26	2.19	2.06
29	2.73	2.69	2.66	2.63	2.60	2.57	2.53	2.49	2.45	2.41	2.33	2.23	2.16	2.03
30	2.70	2.66	2.63	2.60	2.57	2.55	2.51	2.47	2.42	2.39	2.30	2.21	2.13	2.01
31	2.68	2.64	2.61	2.58	2.55	2.52	2.48	2.45	2.40	2.36	2.27	2.18	2.11	1.98
32	2.65	2.62	2.58	2.55	2.53	2.50	2.46	2.42	2.38	2.34	2.25	2.16	2.08	1.96
33	2.63	2.60	2.56	2.53	2.51	2.48	2.44	2.40	2.36	2.32	2.23	2.14	2.06	1.93
34	2.61	2.58	2.54	2.51	2.49	2.46	2.42	2.38	2.34	2.30	2.21	2.12	2.04	1.91
35	2.60	2.56	2.53	2.50	2.47	2.44	2.40	2.36	2.32	2.28	2.19	2.10	2.02	1.89
36	2.58	2.54	2.51	2.48	2.45	2.43	2.38	2.35	2.30	2.26	2.18	2.08	2.00	1.87
37	2.56	2.53	2.49	2.46	2.44	2.41	2.37	2.33	2.28	2.25	2.16	2.06	1.98	1.85
38	2.55	2.51	2.48	2.45	2.42	2.40	2.35	2.32	2.27	2.23	2.14	2.05	1.97	1.84
39	2.54	2.50	2.46	2.43	2.41	2.38	2.34	2.30	2.26	2.22	2.13	2.03	1.95	1.82
40	2.52	2.48	2.45	2.42	2.39	2.37	2.33	2.29	2.24	2.20	2.11	2.02	1.94	1.80
45	2.46	2.43	2.39	2.36	2.34	2.31	2.27	2.23	2.18	2.14	2.05	1.96	1.88	1.74
50	2.42	2.38	2.35	2.32	2.29	2.27	2.22	2.18	2.14	2.10	2.01	1.91	1.83	1.68
60	2.35	2.31	2.28	2.25	2.22	2.20	2.15	2.12	2.07	2.03	1.94	1.84	1.75	1.60
70	2.31	2.27	2.23	2.20	2.18	2.15	2.11	2.07	2.02	1.98	1.89	1.78	1.70	1.54
80	2.27	2.23	2.20	2.17	2.14	2.12	2.07	2.03	1.98	1.94	1.85	1.75	1.66	1.49
90	2.24	2.21	2.17	2.14	2.11	2.09	2.04	2.00	1.96	1.92	1.82	1.72	1.62	1.46
100	2.22	2.19	2.15	2.12	2.09	2.07	2.02	1.98	1.93	1.89	1.80	1.69	1.60	1.43
110	2.21	2.17	2.13	2.10	2.07	2.05	2.00	1.96	1.92	1.88	1.78	1.67	1.57	1.40
120	2.19	2.15	2.12	2.09	2.06	2.03	1.99	1.95	1.90	1.86	1.76	1.66	1.54	1.38
140	2.17	2.13	2.10	2.07	2.04	2.01	1.97	1.93	1.88	1.84	1.74	1.63	1.53	1.35
160	2.15	2.11	2.08	2.05	2.02	1.99	1.95	1.91	1.86	1.82	1.72	1.61	1.51	1.32
180	2.14	2.10	2.07	2.04	2.01	1.98	1.94	1.90	1.85	1.81	1.71	1.60	1.50	1.30
200	2.13	2.09	2.06	2.03	2.00	1.97	1.93	1.89	1.84	1.79	1.69	1.58	1.48	1.28
à	2.04	2.00	1.97	1.93	1.90	1.88	1.83	1.79	1.74	1.70	1.59	1.47	1.36	1.00

TABLE 5:
Percentage points of the studentized range

k = Number of means

df	α	2	3	4	5	6
5	0.05	3.64	4.60	5.22	5.67	6.03
6	0.05	3.46	4.34	4.90	5.3	5.63
7	0.05	3.34	4.16	4.68	5.06	5.36
8	0.05	3.26	4.04	4.53	4.89	5.17
9	0.05	3.2	3.95	4.41	4.76	5.02
10	0.05	3.15	3.88	4.33	4.65	4.91
11	0.05	3.11	3.82	4.26	4.57	4.82.
12	0.05	3.08	3.77	4.20	4.51	4.75
13	0.05	3.06	3.73	4.15	4.45	4.69
14	0.05	3.03	3.70	4.11	4.41	4.64
15	0.05	3.01	3.67	4.08	4.37	4.59

TABLE 6:
Random numbers

1735	6040	2537	5480	9607	7165	8376	7704	6253	8711	6338	0933	3734	3541	8013
3261	8742	2304	9303	7416	0565	5450	4154	6596	8879	6744	0285	0510	8070	3515
9259	5782	4890	8924	1708	8867	1952	1557	4592	8362	4715	3392	4152	1515	1212
4035	7559	8763	7540	8831	4679	2634	9421	4160	7124	3779	4261	4552	2777	2567
0290	4533	3135	6361	9181	8035	8864	8848	1910	6995	9393	3668	6865	0907	2540
1164	4842	2873	6089	9329	7601	5677	7791	5219	7374	6237	5750	0175	5226	9720
5966	3457	8748	0895	4598	8470	4230	6950	9633	5212	6010	3953	5994	7137	1089
3141	9842	8447	7162	3588	0899	1051	1157	7245	1020	0524	6272	9182	8761	3740
1252	8064	3481	4190	1143	6387	7079	2801	0159	1781	0733	7198	9739	7092	3640
6978	4272	1341	7000	7980	2319	2584	5282	5958	1674	4146	3629	7730	9532	5685
1299	0796	7496	7440	4156	6879	4664	2674	0835	5061	7999	2398	7383	5947	3686
9926	0374	0643	8959	8106	3343	8217	6471	2277	4697	1634	2177	8672	2312	5497
3712	7751	4376	7986	1891	8062	1276	7815	0532	1335	7942	1965	0922	8934	7233
2762	5147	0411	1731	3913	8593	7340	0314	9319	2465	0271	4302	6616	5774	2501
8905	2781	3558	6024	6778	6340	3366	0465	9142	4588	4658	2185	3827	5733	9626
7389	6272	8985	9127	4010	6312	3424	4285	1721	7982	4642	6455	8196	1428	7362
9846	6925	9103	1047	6084	4003	9758	9522	2662	0821	9328	8993	5434	4996	5331
0352	6475	9070	4029	6023	3599	3007	8120	0180	8357	8349	3565	8454	6430	8826
7913	5974	9943	9689	9300	3874	3858	7304	5401	2088	9099	9628	3620	3469	6848
8351	9866	8042	6620	7985	5611	3716	4181	8707	8536	6489	4453	8728	2647	6783
2443	9757	3987	0509	8441	7147	8163	4252	2191	6920	6796	2642	2022	2540	3618
3255	7382	7078	8600	6781	4543	6331	4214	4213	5701	8048	7996	9583	4771	5976
2920	7022	5141	0821	9634	4175	5380	5691	3842	4360	2912	8560	8947	8765	6318
3654	9193	4711	3553	9797	7351	6750	3395	5792	4753	1851	1229	0184	1788	3843
5767	3354	3308	0792	0753	3594	0643	8561	8546	3808	4059	8198	7335	2333	1988
7796	2040	1922	3943	1375	0716	0426	5486	8943	1856	3922	5899	6190	9420	0560
7342	5651	9066	4897	6809	5340	8932	0719	9260	3084	9338	3583	5209	2690	9763
3361	6102	7408	6675	0037	0524	5463	1705	0931	0663	7990	8546	4899	2869	6268
3506	2001	6497	0880	2568	3728	3759	0292	2291	9912	5016	7780	7499	1987	9732
6525	7563	2468	9127	3407	8261	5075	6392	7974	1029	8040	6870	4390	7812	7181
1535	4491	9896	2736	5931	7094	3650	2935	0643	8813	6896	0774	3275	6583	9742
1443	5427	3403	1525	7027	9445	0859	8626	2717	5805	3989	7985	6057	8630	8888
3179	9771	9654	1384	6747	9815	7174	3310	5041	5453	7162	2114	8826	9008	2872
1837	2212	1857	0660	4132	1851	1264	9426	0338	6420	3574	1714	7933	9386	5282
5290	9901	0274	5198	0554	9806	1503	2387	7805	7553	9313	4437	1244	7682	1882
0212	9446	3011	2592	1310	4465	7825	9503	3931	3462	3261	9340	8012	1277	8401
7091	4956	2054	2691	8882	2907	6151	7517	3280	9513	4702	2844	4825	5581	9386
4677	1224	3109	8451	3782	4842	2467	5606	8009	9172	1462	5289	9855	5378	8857
5485	8074	4291	1514	0791	3314	1499	9650	8902	8800	5023	9381	0039	5692	1374
4900	0071	4292	3506	1749	1638	8503	7873	0659	1823	5811	6169	5656	7311	7140
8319	9210	2062	5618	6247	9514	0616	5893	5534	1776	4014	4865	7869	8465	9638
9565	2204	3553	7383	5048	7917	0087	2663	2293	3705	8220	2796	2441	3192	8671
1473	7793	8439	7219	4046	5971	0864	7520	5644	7943	1333	8080	5665	5908	0022
9729	3259	3822	1355	2759	5663	7467	0628	2813	4834	7558	6209	7529	5300	4290
3546	9220	9008	2460	3418	5320	7195	3316	3489	8999	7847	1261	0844	4529	9174

STATISTICAL AND INTEREST TABLES

TABLE 7: $(1 + i)^n =$

n \ i	½ % 0.005	¾ % 0.0075	1 % 0.01	1¼ % 0.0125	1½ % 0.015	2 % 0.02
1	1.005 000	1.007 500	1.010 000	1.012 500	1.015 000	1.020 000
2	1.010 025	1.015 056	1.020 100	1.025 156	1.030 225	1.040 400
3	1.015 075	1.022 669	1.030 301	1.037 971	1.045 678	1.061 208
4	1.020 151	1.030 339	1.040 604	1.050 945	1.061 364	1.082 432
5	1.025 251	1.038 067	1.051 010	1.064 082	1.077 284	1.104 081
6	1.030 378	1.045 852	1.061 520	1.077 383	1.093 443	1.126 162
7	1.035 529	1.053 696	1.072 135	1.090 850	1.109 845	1.148 686
8	1.040 707	1.061 599	1.082 857	1.104 486	1.126 493	1.171 659
9	1.045 911	1.069 561	1.093 685	1.118 292	1.143 390	1.195 093
10	1.051 140	1.077 583	1.104 622	1.132 271	1.160 541	1.218 994
11	1.056 396	1.085 664	1.115 668	1.146 424	1.177 949	1.243 374
12	1.061 678	1.093 807	1.126 825	1.160 755	1.195 618	1.268 242
13	1.066 986	1.102 010	1.138 093	1.175 264	1.213 552	1.293 607
14	1.072 321	1.110 276	1.149 474	1.189 955	1.231 756	1.319 479
15	1.077 683	1.118 603	1.160 969	1.204 829	1.250 232	1.345 868
16	1.083 071	1.126 992	1.172 579	1.219 890	1.268 986	1.372 786
17	1.088 487	1.135 445	1.184 304	1.235 138	1.288 020	1.400 241
18	1.093 929	1.143 960	1.196 147	1.250 577	1.307 341	1.428 246
19	1.099 399	1.152 540	1.208 109	1.266 210	1.326 951	1.456 811
20	1.104 896	1.161 184	1.220 190	1.282 037	1.346 855	1.485 947
21	1.110 420	1.169 893	1.232 392	1.298 063	1.367 058	1.515 666
22	1.115 972	1.178 667	1.244 716	1.314 288	1.387 564	1.545 980
23	1.121 552	1.187 507	1.257 163	1.330 717	1.408 377	1.576 899
24	1.127 160	1.196 414	1.269 735	1.347 351	1.429 503	1.608 437
25	1.132 796	1.205 387	1.282 432	1.364 193	1.450 945	1.640 606
26	1.138 460	1.214 427	1.295 256	1.381 245	1.472 710	1.673 418
27	1.144 152	1.223 535	1.308 209	1.398 511	1.494 800	1.706 886
28	1.149 873	1.232 712	1.321 291	1.415 992	1.517 222	1.741 024
29	1.155 622	1.241 957	1.334 504	1.433 692	1.539 981	1.775 845
30	1.161 400	1.251 272	1.347 849	1.451 613	1.563 080	1.811 362
31	1.167 207	1.260 656	1.361 327	1.469 759	1.586 526	1.847 589
32	1.173 043	1.270 111	1.374 941	1.488 131	1.610 324	1.884 541
33	1.178 908	1.279 637	1.388 690	1.506 732	1.634 479	1.922 231
34	1.184 803	1.289 234	1.402 577	1.525 566	1.658 996	1.960 676
35	1.190 727	1.298 904	1.416 603	1.544 636	1.683 881	1.999 890
36	1.196 681	1.308 645	1.430 769	1.563 944	1.709 140	2.039 887
37	1.202 664	1.318 460	1.445 076	1.583 493	1.734 777	2.080 685
38	1.208 677	1.328 349	1.459 527	1.603 287	1.760 798	2.122 299
39	1.214 721	1.338 311	1.474 123	1.623 328	1.787 210	2.164 745
40	1.220 794	1.348 349	1.488 864	1.643 619	1.814 018	2.208 040
41	1.226 898	1.358 461	1.503 752	1.664 165	1.841 229	2.252 200
42	1.233 033	1.368 650	1.518 790	1.684 967	1.868 847	2.297 244
43	1.239 198	1.378 915	1.533 978	1.706 029	1.896 880	2.343 189
44	1.245 394	1.389 256	1.549 318	1.727 354	1.925 333	2.390 053
45	1.251 621	1.399 676	1.564 811	1.748 946	1.954 213	2.437 854
46	1.257 879	1.410 173	1.580 459	1.770 808	1.983 526	2.486 611
47	1.264 168	1.420 750	1.596 263	1.792 943	2.013 279	2.536 344
48	1.270 489	1.431 405	1.612 226	1.815 355	2.043 478	2.587 070
49	1.276 842	1.442 141	1.628 348	1.838 047	2.074 130	2.638 812
50	1.283 226	1.452 957	1.644 632	1.861 022	2.105 242	2.691 588
55	1.315 629	1.508 266	1.728 525	1.980 281	2.267 944	2.971 731
60	1.348 850	1.565 681	1.816 697	2.107 181	2.443 220	3.281 031
65	1.382 910	1.625 281	1.909 366	2.242 214	2.632 042	3.622 523
70	1.417 831	1.687 151	2.006 763	2.385 900	2.835 456	3.999 558
75	1.453 633	1.751 375	2.109 128	2.538 794	3.054 592	4.415 835
80	1.490 339	1.818 044	2.216 715	2.701 485	3.290 663	4.875 439
85	1.527 971	1.887 251	2.329 790	2.874 602	3.544 978	5.382 879
90	1.566 555	1.959 092	2.448 633	3.058 813	3.818 949	5.943 133
95	1.606 112	2.033 669	2.573 538	3.254 828	4.114 092	6.561 699
100	1.646 668	2.111 084	2.704 814	3.463 404	4.432 046	7.244 646

TABLE 7 (continued): $(1 + i)^n =$

n \ i	2½ % 0.025	3 % 0.03	3½ % 0.035	4 % 0.04	4½ % 0.045	5 % 0.05
1	1.025 000	1.030 000	1.035 000	1.040 000	1.045 000	1.050 000
2	1.050 625	1.060 900	1.071 225	1.081 600	1.092 025	1.102 500
3	1.076 891	1.092 727	1.108 718	1.124 864	1.141 166	1.157 625
4	1.103 813	1.125 509	1.147 523	1.169 859	1.192 519	1.215 506
5	1.131 408	1.159 274	1.187 686	1.216 653	1.246 182	1.276 282
6	1.159 693	1.194 052	1.229 255	1.265 319	1.302 260	1.340 096
7	1.188 686	1.229 874	1.272 279	1.315 932	1.360 862	1.407 100
8	1.218 403	1.266 770	1.316 809	1.368 569	1.422 101	1.477 455
9	1.248 863	1.304 773	1.362 897	1.423 312	1.486 095	1.551 328
10	1.280 085	1.343 916	1.410 599	1.480 244	1.552 969	1.628 895
11	1.312 087	1.384 234	1.459 970	1.539 454	1.622 853	1.710 339
12	1.344 889	1.425 761	1.511 069	1.601 032	1.695 881	1.795 856
13	1.378 511	1.468 534	1.563 956	1.665 074	1.772 196	1.885 649
14	1.412 974	1.512 590	1.618 695	1.731 686	1.851 945	1.979 932
15	1.448 298	1.557 967	1.675 349	1.800 944	1.935 282	2.078 928
16	1.484 506	1.604 706	1.733 986	1.872 981	2.022 370	2.182 875
17	1.521 618	1.652 848	1.794 676	1.947 900	2.113 377	2.292 018
18	1.559 659	1.702 433	1.857 489	2.025 817	2.208 479	2.406 619
19	1.598 650	1.753 506	1.922 501	2.106 849	2.307 860	2.526 950
20	1.638 616	1.806 111	1.989 789	2.191 123	2.411 714	2.653 298
21	1.679 582	1.860 295	2.059 431	2.278 768	2.520 241	2.785 963
22	1.721 571	1.916 103	2.131 512	2.369 919	2.633 652	2.925 261
23	1.764 611	1.973 587	2.206 114	2.464 716	2.752 166	3.071 524
24	1.808 726	2.032 794	2.283 328	2.563 304	2.876 014	3.225 100
25	1.853 944	2.093 778	2.363 245	2.665 836	3.005 434	3.386 355
26	1.900 293	2.156 591	2.445 959	2.772 470	3.140 679	3.555 673
27	1.947 800	2.221 289	2.531 567	2.883 369	3.282 010	3.733 456
28	1.996 495	2.287 928	2.620 172	2.998 703	3.429 700	3.920 129
29	2.046 407	2.356 566	2.711 878	3.118 651	3.584 036	4.116 136
30	2.097 568	2.427 262	2.806 794	3.243 398	3.745 318	4.321 942
31	2.150 007	2.500 080	2.905 031	3.373 133	3.913 857	4.538 039
32	2.203 757	2.575 083	3.006 708	3.508 059	4.089 981	4.764 941
33	2.258 851	2.652 335	3.111 942	3.648 381	4.274 030	5.003 189
34	2.315 322	2.731 905	3.220 860	3.794 316	4.466 362	5.253 348
35	2.373 205	2.813 862	3.333 590	3.946 089	4.667 348	5.516 015
36	2.432 535	2.898 278	3.450 266	4.103 933	4.877 378	5.791 816
37	2.493 349	2.985 227	3.571 025	4.268 090	5.096 860	6.081 407
38	2.555 682	3.074 783	3.696 011	4.438 813	5.326 219	6.385 477
39	2.619 574	3.167 027	3.825 372	4.616 366	5.565 899	6.704 751
40	2.685 064	3.262 038	3.959 260	4.801 021	5.816 365	7.039 989
41	2.752 190	3.359 899	4.097 834	4.993 061	6.078 101	7.391 988
42	2.820 995	3.460 696	4.241 258	5.192 784	6.351 615	7.761 588
43	2.891 520	3.564 517	4.389 702	5.400 495	6.637 438	8.149 667
44	2.963 808	3.671 452	4.543 342	5.616 515	6.936 123	8.557 150
45	3.037 903	3.781 596	4.702 359	5.841 176	7.248 248	8.985 008
46	3.113 851	3.895 044	4.866 941	6.074 823	7.574 420	9.434 258
47	3.191 697	4.011 895	5.037 284	6.317 816	7.915 268	9.905 971
48	3.271 490	4.132 252	5.213 589	6.570 528	8.271 456	10.401 270
49	3.353 277	4.256 219	5.396 065	6.833 349	8.643 671	10.921 333
50	3.437 109	4.383 906	5.584 927	7.106 683	9.032 636	11.467 400
55	3.888 773	5.082 149	6.633 141	8.646 367	11.256 308	14.635 631
60	4.399 790	5.891 603	7.878 091	10.519 627	14.027 408	18.679 186
65	4.977 958	6.829 983	9.356 701	12.798 735	17.480 702	23.839 901
70	5.632 103	7.917 822	11.112 825	15.571 618	21.784 136	30.426 426
75	6.372 207	9.178 926	13.198 550	18.945 255	27.146 996	38.832 686
80	7.209 568	10.640 891	15.675 738	23.049 799	33.830 096	49.561 441
85	8.156 964	12.335 709	18.617 859	28.043 605	42.158 455	63.254 353
90	9.228 856	14.300 467	22.112 176	34.119 333	52.537 105	80.730 365
95	10.441 604	16.578 161	26.262 329	41.511 386	65.470 792	103.034 676
100	11.813 716	19.218 632	31.191 408	50.504 948	81.588 518	131.501 258

STATISTICAL AND INTEREST TABLES

TABLE 7 (continued): $(1 + i)^n =$

n \ i	5½ % 0.055	6 % 0.06	6½ % 0.065	7 % 0.07	8 % 0.08	9 % 0.09
1	1.055 000	1.060 000	1.065 000	1.070 000	1.080 000	1.090 000
2	1.113 025	1.123 600	1.134 225	1.144 900	1.166 400	1.188 100
3	1.174 241	1.191 016	1.207 950	1.225 043	1.259 712	1.295 029
4	1.238 825	1.262 477	1.286 466	1.310 796	1.360 489	1.411 582
5	1.306 960	1.338 226	1.370 087	1.402 552	1.469 328	1.538 624
6	1.378 843	1.418 519	1.459 142	1.500 730	1.586 874	1.677 100
7	1.454 679	1.503 630	1.553 987	1.605 781	1.713 824	1.828 039
8	1.534 687	1.593 848	1.654 996	1.718 186	1.850 930	1.992 563
9	1.619 094	1.689 479	1.762 570	1.838 459	1.999 005	2.171 893
10	1.708 144	1.790 848	1.877 137	1.967 151	2.158 925	2.367 364
11	1.802 092	1.898 299	1.999 151	2.104 852	2.331 639	2.580 426
12	1.901 207	2.012 196	2.129 096	2.252 192	2.518 170	2.812 665
13	2.005 774	2.132 928	2.267 487	2.409 845	2.719 624	3.065 805
14	2.116 091	2.260 904	2.414 874	2.578 534	2.937 194	3.341 727
15	2.232 476	2.396 558	2.571 841	2.759 032	3.172 169	3.642 482
16	2.355 263	2.540 352	2.739 011	2.952 164	3.425 943	3.970 306
17	2.484 802	2.692 773	2.917 046	3.158 815	3.700 018	4.327 633
18	2.621 466	2.854 339	3.106 654	3.379 932	3.996 019	4.717 120
19	2.765 647	3.025 600	3.308 587	3.616 528	4.315 701	5.141 661
20	2.917 757	3.207 135	3.523 645	3.869 684	4.660 957	5.604 411
21	3.078 234	3.399 564	3.752 682	4.140 562	5.033 834	6.108 808
22	3.247 537	3.603 537	3.996 606	4.430 402	5.436 540	6.658 600
23	3.426 152	3.819 750	4.256 386	4.740 530	5.871 464	7.257 874
24	3.614 590	4.048 935	4.533 051	5.072 367	6.341 181	7.911 083
25	3.813 392	4.291 871	4.827 699	5.427 433	6.848 475	8.623 081
26	4.023 129	4.549 383	5.141 500	5.807 353	7.396 353	9.399 158
27	4.244 401	4.822 346	5.475 697	6.213 868	7.988 061	10.245 082
28	4.477 843	5.111 687	5.831 617	6.648 838	8.627 106	11.167 140
29	4.724 124	5.418 388	6.210 672	7.114 257	9.317 275	12.172 182
30	4.983 951	5.743 491	6.614 366	7.612 255	10.062 657	13.267 678
31	5.258 069	6.088 101	7.044 300	8.145 113	10.867 669	14.461 770
32	5.547 262	6.453 387	7.502 179	8.715 271	11.737 083	15.763 329
33	5.852 362	6.840 590	7.989 821	9.325 340	12.676 050	17.182 028
34	6.174 242	7.251 025	8.509 159	9.978 114	13.690 134	18.728 411
35	6.513 825	7.686 087	9.062 255	10.676 581	14.785 344	20.413 968
36	6.872 085	8.147 252	9.651 301	11.423 942	15.968 172	22.251 225
37	7.250 050	8.636 087	10.278 636	12.223 618	17.245 626	24.253 835
38	7.648 803	9.154 252	10.946 747	13.079 271	18.625 276	26.436 680
39	8.069 487	9.703 507	11.658 286	13.994 820	20.115 298	28.815 982
40	8.513 309	10.285 718	12.416 075	14.974 458	21.724 521	31.409 420
41	8.981 541	10.902 861	13.223 119	16.022 670	23.462 483	34.236 268
42	9.475 525	11.557 033	14.082 622	17.144 257	25.339 482	37.317 532
43	9.996 679	12.250 455	14.997 993	18.344 355	27.366 640	40.676 110
44	10.546 497	12.985 482	15.972 862	19.628 460	29.555 972	44.336 960
45	11.126 554	13.764 611	17.011 098	21.002 452	31.920 449	48.327 286
46	11.738 515	14.590 487	18.116 820	22.472 623	34.474 085	52.676 742
47	12.384 133	15.465 917	19.294 413	24.045 707	37.232 012	54.417 649
48	13.065 260	16.393 872	20.548 550	25.728 907	40.210 573	62.585 237
49	13.783 849	17.377 504	21.884 205	27.529 930	43.427 419	68.217 908
50	14.541 961	18.420 154	23.306 679	29.457 025	46.901 613	74.357 520
55	19.005 762	24.650 322	31.932 170	41.315 001	68.913 856	114.408 262
60	24.839 770	32.987 691	43.749 840	57.946 427	101.257 064	176.031 292
65	32.464 587	44.144 972	59.941 072	81.272 861	148.779 847	270.845 963
70	42.429 916	59.075 930	82.124 463	113.989 392	218.606 406	416.730 086
75	55.454 204	79.056 921	112.517 632	159.876 019	321.204 530	641.190 893
80	72.476 426	105.795 993	154.158 907	224.234 388	471.954 834	986.551 668
85	94.723 791	141.578 904	211.211 062	314.500 328	693.456 489	1 517.932 029
90	123.800 206	189.464 511	289.377 460	441.102 980	1 018.915 089	2 335.526 582
95	161.801 918	253.546 255	396.472 198	618.669 748	1 497.120 549	3 593.497 147
100	211.468 636	339.302 084	543.201 271	867.716 326	2 199.761 256	5 529.040 792

TABLE 7 (continued): $(1 + i)^n$

n \ i	10 % 0.1	11 % 0.11	12 % 0.12	13 % 0.13	14 % 0.14	15 % 0.15
1	1.100 000	1.110 000	1.120 000	1.130 000	1.140 000	1.150 000
2	1.210 000	1.232 100	1.254 400	1.276 900	1.299 600	1.322 500
3	1.331 000	1.367 631	1.404 928	1.442 897	1.481 544	1.520 875
4	1.464 100	1.518 070	1.573 519	1.630 474	1.688 960	1.749 006
5	1.610 510	1.685 058	1.762 342	1.842 435	1.925 415	2.011 357
6	1.771 561	1.870 415	1.973 823	2.081 952	2.194 973	2.313 061
7	1.948 717	2.076 160	2.210 681	2.352 605	2.502 269	2.660 020
8	2.143 589	2.304 538	2.475 963	2.658 444	2.852 586	3.059 023
9	2.357 948	2.558 037	2.773 079	3.004 042	3.251 949	3.517 876
10	2.593 742	2.839 421	3.105 848	3.394 567	3.707 221	4.045 558
11	2.853 117	3.151 757	3.478 550	3.835 861	4.226 232	4.652 391
12	3.138 428	3.498 451	3.895 976	4.334 523	4.817 905	5.350 250
13	3.452 271	3.883 280	4.363 493	4.898 011	5.492 411	6.152 788
14	3.797 498	4.310 441	4.887 112	5.534 753	6.261 349	7.075 706
15	4.177 248	4.784 589	5.473 566	6.254 270	7.137 938	8.137 062
16	4.594 973	5.310 894	6.130 394	7.067 326	8.137 249	9.357 621
17	5.054 470	5.895 093	6.866 041	7.986 078	9.276 464	10.761 264
18	5.559 917	6.543 553	7.689 966	9.024 268	10.575 169	12.375 454
19	6.115 909	7.263 344	8.612 762	10.197 423	12.055 693	14.231 772
20	6.727 500	8.062 312	9.646 293	11.523 088	13.743 490	16.366 537
21	7.400 250	8.949 166	10.803 848	13.021 089	15.667 578	18.821 518
22	8.140 275	9.933 574	12.100 310	14.713 831	17.861 039	21.644 746
23	8.954 302	11.026 267	13.552 347	16.626 629	20.361 585	24.891 458
24	9.849 733	12.239 157	15.178 629	18.788 091	23.212 207	28.625 176
25	10.834 706	13.585 464	17.000 064	21.230 542	26.461 916	32.918 953
26	11.918 177	15.079 865	19.040 072	23.990 513	30.166 584	37.856 796
27	13.109 994	16.738 650	21.324 881	27.109 279	34.389 906	43.535 315
28	14.420 994	18.579 901	23.883 866	30.633 486	39.204 493	50.065 612
29	15.863 093	20.623 691	26.749 930	34.615 839	44.693 122	57.575 454
30	17.449 402	22.892 297	29.959 922	39.115 898	50.950 159	66.211 772
31	19.194 342	25.410 449	33.555 113	44.200 965	58.083 181	76.143 538
32	21.113 777	28.205 599	37.581 726	49.947 090	66.214 826	87.565 068
33	23.225 154	31.308 214	42.091 533	56.440 212	75.484 902	100.699 829
34	25.547 670	34.752 118	47.142 517	63.777 439	86.052 788	115.804 803
35	28.102 437	38.574 851	52.799 620	72.068 506	98.100 178	133.175 523
36	30.912 681	42.818 085	59.135 574	81.437 412	111.834 203	153.151 852
37	34.003 949	47.528 074	66.231 843	92.024 276	127.490 992	176.124 630
38	37.404 343	52.756 162	74.179 664	103.987 432	145.339 731	202.543 324
39	41.144 778	58.559 340	83.081 224	117.505 798	165.687 293	232.924 823
40	45.259 256	65.000 867	93.050 970	132.781 552	188.883 514	267.863 546
41	49.785 181	72.150 963	104.217 087	150.043 153	215.327 206	308.043 078
42	54.763 699	80.087 569	116.723 137	169.548 763	245.473 015	354.249 540
43	60.240 069	88.897 201	130.729 914	191.590 103	279.839 237	407.386 971
44	66.264 076	98.675 893	146.417 503	216.496 816	319.016 730	468.495 017
45	72.890 484	109.530 242	163.987 604	244.641 402	363.679 072	538.769 269
46	80.179 532	121.578 568	183.666 116	276.444 784	414.594 142	619.584 659
47	88.197 485	134.952 211	205.706 050	312.382 606	472.637 322	712.522 358
48	97.017 234	149.796 954	230.390 776	352.992 345	538.806 547	819.400 712
49	106.718 957	166.274 619	258.037 669	398.881 350	614.239 464	942.310 819
50	117.390 853	184.564 827	289.002 190	450.735 925	700.232 988	1 083.657 442
55	189.059 142	311.002 466	509.320 606	830.451 725	1 348.238 807	2 179.622 184
60	304.481 640	524.057 242	897.596 933	1 530.053 473	2 595.918 660	4 383.998 746
65	490.370 725	883.066 930	1 581.872 491	2 819.024 345	4 998.219 642	8 817.787 387
70	789.746 957	1 488.019 132	2 787.799 828	5 193.869 624	9 623.644 985	17 735.720 039
75	1 271.895 371	2 507.398 773	4 913.055 841	9 569.368 113	18 529.506 390	35 672.867 976
80	2 048.400 215	4 225.112 750	8 658.483 100	17 630.940 454	35 676.981 807	71 750.879 401
85	3 298.969 030	7 119.560 696	15 259.205 681	32 483.864 937	68 692.981 028	144 316.646 994
90	5 313.022 612	11 996.873 812	26 891.934 223	59 849.415 520	132 262.467 379	290 272.325 206
95	8 556.676 047	20 215.430 053	47 392.776 624	110 268.668 614	254 660.083 396	583 841.327 636
100	13 780.612 340	34 064.175 270	83 522.265 727	203 162.874 228	490 326.238 126	1 174 313.450 700

STATISTICAL AND INTEREST TABLES

TABLE 8: $s_{\overline{n}|}\ i = \dfrac{(1+i)^n - 1)}{i}$

n \ i	½ % 0.005	¾ % 0.0075	1 % 0.01	1¼ % 0.0125	1½ % 0.015	2 % 0.02
1	1.000 000	1.000 000	1.000 000	1.000 000	1.000 000	1.000 000
2	2.005 000	2.007 500	2.010 000	2.012 500	2.015 000	2.020 000
3	3.015 025	3.022 556	3.030 100	3.037 656	3.045 225	3.060 400
4	4.030 100	4.045 225	4.060 401	4.075 627	4.090 903	4.121 608
5	5.050 251	5.075 565	5.101 005	5.126 572	5.152 267	5.204 040
6	6.075 502	6.113 631	6.152 015	6.190 654	6.229 551	6.308 121
7	7.105 879	7.159 484	7.213 535	7.268 038	7.322 994	7.434 283
8	8.141 409	8.213 180	8.285 671	8.358 888	8.432 839	8.582 969
9	9.182 116	9.274 779	9.368 527	9.463 374	9.559 332	9.754 628
10	10.228 026	10.344 339	10.462 213	10.581 666	10.702 722	10.949 721
11	11.279 167	11.421 922	11.566 835	11.713 937	11.863 262	12.168 715
12	12.335 562	12.507 586	12.682 503	12.860 361	13.041 211	13.412 090
13	13.397 240	13.601 393	13.809 328	14.021 116	14.236 830	14.680 332
14	14.464 226	14.703 404	14.947 421	15.196 380	15.450 382	15.973 938
15	15.536 548	15.813 679	16.096 896	16.386 335	16.682 138	17.293 417
16	16.614 230	16.932 282	17.257 864	17.591 164	17.932 370	18.639 285
17	17.697 301	18.059 274	18.430 443	18.811 053	19.201 355	20.012 071
18	18.785 788	19.194 718	19.614 748	20.046 192	20.489 376	21.412 312
19	19.879 717	20.338 679	20.810 895	21.296 769	21.796 716	22.840 559
20	20.979 115	21.491 219	22.019 004	22.562 979	23.123 667	24.297 370
21	22.084 011	22.652 403	23.239 194	23.845 016	24.470 522	25.783 317
22	23.194 431	23.822 296	24.471 586	25.143 078	25.837 580	27.298 984
23	24.310 403	25.000 963	25.716 302	26.457 367	27.225 144	28.844 963
24	25.431 955	26.188 471	26.973 465	27.788 084	28.633 521	30.421 862
25	26.559 115	27.384 884	28.243 200	29.135 435	30.063 024	32.030 300
26	27.691 911	28.590 271	29.525 631	30.499 628	31.513 969	33.670 906
27	28.830 370	29.804 698	30.820 888	31.880 873	32.986 678	35.344 324
28	29.974 522	31.028 233	32.129 097	33.279 384	34.481 479	37.051 210
29	31.124 395	32.260 945	33.450 388	34.695 377	35.998 701	38.792 235
30	32.280 017	33.502 902	34.784 892	36.129 069	37.538 681	40.568 079
31	33.441 417	34.754 174	36.132 740	37.580 682	39.101 762	42.379 441
32	34.608 624	36.014 830	37.494 068	39.050 441	40.688 288	44.227 030
33	35.781 667	37.284 941	38.869 009	40.538 571	42.298 612	46.111 570
34	36.960 575	38.564 578	40.257 699	42.045 303	43.933 092	48.033 802
35	38.145 378	39.853 813	41.660 276	43.570 870	45.592 088	49.994 478
36	39.336 105	41.152 716	43.076 878	45.115 505	42.275 969	51.994 367
37	40.532 785	42.461 361	44.507 647	46.679 449	48.985 109	54.034 255
38	41.735 449	43.779 822	45.952 724	48.262 942	50.719 885	56.114 940
39	42.944 127	45.108 170	47.412 251	49.866 229	52.480 684	58.237 238
40	44.158 847	46.446 482	48.886 373	51.489 557	54.267 894	60.401 983
41	45.379 642	47.794 830	50.375 237	53.133 177	56.081 912	62.610 023
42	46.606 540	49.153 291	51.878 989	54.797 341	57.923 141	64.862 223
43	47.839 572	50.521 941	53.397 779	56.482 308	59.791 988	67.159 468
44	49.078 770	51.900 856	54.931 757	58.188 337	61.688 868	69.502 657
45	50.324 164	53.290 112	56.481 075	59.915 691	63.614 201	71.892 710
46	51.575 785	54.689 788	58.045 885	61.664 637	65.568 414	74.330 564
47	52.833 664	56.099 961	59.626 344	63.435 445	67.551 940	76.817 176
48	54.097 832	57.520 711	61.222 608	65.228 388	69.565 219	79.353 519
49	55.368 321	58.952 116	62.834 834	67.043 743	71.608 698	81.940 590
50	56.645 163	60.394 275	64.463 182	68.881 790	73.682 828	84.579 401
55	63.125 775	67.768 834	72.852 457	78.422 456	84.529 599	98.586 534
60	69.770 031	75.424 137	81.669 670	88.574 508	96.214 652	114.051 539
65	76.582 062	83.370 852	90.936 649	99.377 125	108.802 772	131.126 155
70	83.566 105	91.620 073	100.676 337	110.871 998	122.363 753	149.977 911
75	90.726 505	100.183 314	110.912 847	123.103 486	136.972 781	170.791 773
80	98.067 714	109.072 531	121.671 522	136.118 795	152.710 852	193.771 958
85	105.594 297	118.300 130	132.978 997	149.968 153	169.665 226	219.143 939
90	113.310 936	127.878 995	144.863 267	164.705 008	187.929 900	247.156 656
95	121.222 430	137.822 495	157.353 755	180.386 232	207.606 142	278.084 960
100	129.333 698	148.144 512	170.481 383	197.072 342	228.803 043	312.232 306

TABLE 8 (continued): $s_{\overline{n}|}\ i = \dfrac{(1+i)^n - 1}{i}$

n \ i	2½ % 0.025	3 % 0.03	3½ % 0.035	4 % 0.04	4½ % 0.045	5 % 0.05
1	1.000 000	1.000 000	1.000 000	1.000 000	1.000 000	1.000 000
2	2.025 000	2.030 000	2.035 000	2.040 000	2.045 000	2.050 000
3	3.075 625	3.090 900	3.106 225	3.121 600	3.137 025	3.152 500
4	4.152 516	4.183 627	4.214 943	4.246 464	4.278 191	4.310 125
5	5.256 329	5.309 136	5.362 466	5.416 323	5.470 710	5.525 631
6	6.387 737	6.468 410	6.550 152	6.632 975	6.716 892	6.801 913
7	7.547 430	7.662 462	7.779 408	7.898 294	8.019 152	8.142 008
8	8.736 116	8.892 336	9.051 687	9.214 226	9.380 014	9.549 109
9	9.954 519	10.159 106	10.368 496	10.582 795	10.802 114	11.026 564
10	11.203 382	11.463 879	11.731 393	12.006 107	12.288 209	12.577 893
11	12.483 466	12.807 796	13.141 992	13.486 351	13.841 179	14.206 787
12	13.795 553	14.192 030	14.601 962	15.025 805	15.464 032	15.917 127
13	15.140 442	15.617 790	16.113 030	16.626 838	17.159 913	17.712 983
14	16.518 953	17.086 324	17.676 986	18.291 911	18.932 109	19.598 632
15	17.931 927	18.598 914	19.295 681	20.023 588	20.784 054	21.578 564
16	19.380 225	20.156 881	20.971 030	21.824 531	22.719 337	23.657 492
17	20.864 730	21.761 588	22.705 016	23.697 512	24.741 707	25.840 366
18	22.386 349	23.414 435	24.499 691	25.645 413	26.855 084	28.132 385
19	23.946 007	25.116 868	26.357 180	27.671 229	29.063 562	30.539 004
20	25.544 658	26.870 374	28.279 682	29.778 079	31.371 423	33.065 954
21	27.183 274	28.676 486	30.269 471	31.969 202	33.783 137	35.719 252
22	28.862 856	30.536 780	32.328 902	34.247 970	36.303 378	38.505 214
23	30.584 427	32.452 884	34.460 414	36.617 889	38.937 030	41.430 475
24	32.349 038	34.426 470	36.666 528	39.082 604	41.689 196	44.501 999
25	34.157 764	36.459 264	38.949 857	41.645 908	44.565 210	47.727 099
26	36.011 708	38.553 042	41.313 102	44.311 745	47.570 645	51.113 454
27	37.912 001	40.709 634	43.759 060	47.084 214	50.711 324	54.669 126
28	39.859 801	49.930 923	46.290 627	49.967 583	53.993 333	58.402 583
29	41.856 296	45.218 850	48.910 799	52.966 286	57.423 033	62.322 712
30	43.902 703	47.575 416	51.622 677	56.084 938	61.007 070	66.438 848
31	46.000 271	50.002 678	54.429 471	59.328 335	64.752 388	70.760 790
32	48.150 278	52.502 759	57.334 502	62.701 469	68.666 245	75.298 829
33	50.354 034	55.077 841	60.341 210	66.209 527	72.756 226	80.063 771
34	52.612 885	57.730 177	63.453 152	69.857 909	77.030 256	85.066 959
35	54.928 207	60.462 082	66.674 013	73.652 225	81.496 618	90.320 307
36	57.301 413	63.275 944	70.007 603	77.598 314	86.163 966	95.836 323
37	59.733 948	66.174 223	73.457 869	81.702 246	91.041 344	101.628 139
38	62.227 297	69.159 449	77.028 895	85.970 336	96.138 205	107.709 546
39	64.782 979	72.234 233	80.724 906	90.409 150	101.464 424	114.095 023
40	67.402 554	75.401 260	84.550 278	95.025 516	107.030 323	120.799 774
41	70.087 617	78.663 298	88.509 537	99.826 536	112.846 688	127.839 763
42	72.839 808	82.023 196	92.607 371	104.819 598	118.924 789	135.231 751
43	75.660 803	85.483 892	96.848 629	110.012 382	125.276 404	142.993 339
44	78.552 323	89.048 409	101.238 331	115.412 877	131.913 842	151.143 006
45	81.516 131	92.719 861	105.781 673	121.029 392	138.849 965	157.700 156
46	84.554 034	96.501 457	110.484 031	126.870 568	146.098 214	168.685 164
47	87.667 885	100.396 501	115.350 973	132.945 390	153.672 633	178.119 422
48	90.859 582	104.408 396	120.388 257	139.263 206	161.587 902	188.025 393
49	94.131 072	108.540 648	125.601 846	145.833 734	169.859 357	198.426 663
50	97.484 349	112.796 867	130.997 910	152.667 084	178.503 028	209.347 996
55	115.550 921	136.071 620	160.946 890	191.159 173	227.917 959	272.712 618
60	135.991 590	163.053 437	196.516 883	237.990 685	289.497 954	353.583 718
65	159.118 330	194.332 758	238.762 876	294.968 380	366.237 831	456.798 011
70	185.284 114	230.594 064	288.937 865	364.290 459	461.869 680	588.528 511
75	214.888 297	272.630 856	348.530 011	448.631 367	581.044 362	756.653 718
80	248.382 713	321.363 019	419.306 787	551.244 977	729.557 699	971.228 821
85	286.278 570	377.856 952	503.367 394	676.090 123	914.632 336	1 245.087 069
90	329.154 253	443.348 904	603.205 027	827.983 334	1 145.269 007	1 594.607 301
95	377.664 154	519.272 026	721.780 816	1 012.784 648	1 432.684 259	2 040.693 529
100	432.548 654	607.287 733	862.611 657	1 237.623 705	1 790.855 956	2 610.025 157

STATISTICAL AND INTEREST TABLES

TABLE 8 (continued): $s_{\overline{n}|}$ $i = \dfrac{(1+i)^n - 1}{i}$

i \ n	5½ % 0.055	6 % 0.06	6½ % 0.065	7 % 0.07	8 % 0.08	9 % 0.09
1	1.000 000	1.000 000	1.000 000	1.000 000	1.000 000	1.000 000
2	2.055 000	2.060 000	2.065 000	2.070 000	2.080 000	2.090 000
3	3.168 025	3.183 600	3.199 225	3.214 900	3.246 400	3.278 100
4	4.342 266	4.374 616	4.407 175	4.439 943	4.506 112	4.573 129
5	5.581 091	5.637 093	5.693 641	5.750 739	5.866 601	5.984 711
6	6.888 051	6.975 319	7.063 728	7.153 291	7.335 929	7.523 335
7	8.266 894	8.393 838	8.522 870	8.654 021	8.922 803	9.200 435
8	9.721 573	9.897 468	10.076 856	10.259 803	10.636 628	11.028 474
9	11.256 260	11.491 316	11.731 852	11.977 989	12.487 558	13.021 036
10	12.875 354	13.180 795	13.494 423	13.816 448	14.486 562	15.192 930
11	14.583 498	14.971 643	15.371 560	15.783 599	16.645 487	17.560 293
12	16.385 591	16.869 941	17.370 711	17.888 451	18.977 126	20.140 720
13	18.286 798	18.882 138	19.499 808	20.140 643	21.495 297	22.953 385
14	20.292 572	21.015 066	21.767 295	22.550 488	24.214 920	26.019 189
15	22.408 663	23.275 970	24.182 169	25.129 022	27.152 114	29.360 916
16	24.641 140	25.672 528	26.754 010	27.888 054	30.324 283	33.003 399
17	26.996 403	28.212 880	29.493 021	30.840 217	33.750 226	36.973 705
18	29.481 205	30.905 653	32.410 067	33.999 033	37.450 244	41.301 338
19	32.102 671	33.759 992	35.516 722	37.378 965	41.446 263	46.018 458
20	34.868 318	36.785 591	38.825 309	40.995 492	45.761 964	51.160 120
21	37.786 076	39.992 727	42.348 954	44.865 177	50.422 921	56.764 530
22	40.864 310	43.392 290	46.101 636	49.005 739	55.456 755	62.873 338
23	44.111 847	46.995 828	50.098 242	53.436 141	60.893 296	69.531 939
24	47.537 998	50.815 577	54.354 628	58.176 671	66.764 759	76.789 813
25	51.152 588	54.864 512	58.887 679	63.249 038	73.105 940	84.700 896
26	54.965 981	59.156 383	63.715 378	68.676 470	79.954 415	93.323 977
27	58.989 109	63.705 766	68.856 877	74.483 823	87.350 768	102.723 135
28	63.233 510	68.528 112	74.332 574	80.697 691	95.338 830	112.968 217
29	67.711 354	73.639 798	80.164 192	87.346 529	103.965 936	124.135 356
30	72.435 478	79.058 186	86.374 864	94.460 786	113.283 211	136.307 539
31	77.419 429	84.801 677	92.989 230	102.073 041	123.345 868	149.575 217
32	82.677 498	90.889 778	100.033 530	110.218 154	134.213 537	164.036 987
33	88.224 760	97.343 165	107.535 710	118.933 425	145.950 620	179.800 315
34	94.077 122	104.183 755	115.525 531	128.258 765	158.626 670	196.982 344
35	100.251 364	111.434 780	124.034 690	138.236 878	172.316 804	215.710 755
36	106.765 189	119.120 867	133.096 945	148.913 460	187.102 148	236.124 723
37	113.637 274	127.268 119	142.748 247	160.337 402	203.070 320	258.375 948
38	120.887 324	135.904 206	153.026 883	172.561 020	220.315 945	282.629 783
39	128.536 127	145.058 458	163.973 630	185.640 292	238.941 221	309.066 463
40	136.605 614	154.761 966	175.631 916	199.635 112	259.056 519	337.882 445
41	145.118 923	165.047 684	188.047 990	214.609 570	280.781 040	369.291 865
42	154.100 464	175.950 545	201.271 110	230.632 240	304.243 523	403.528 133
43	163.575 989	187.507 577	215.353 732	247.776 496	329.583 005	440.845 665
44	173.572 669	199.758 032	230.351 725	266.120 851	356.949 646	481.521 775
45	184.119 165	212.743 514	246.324 587	285.749 311	386.505 617	525.858 734
46	195.245 719	226.508 125	263.335 685	306.751 763	418.426 067	574.186 021
47	206.984 234	241.098 612	281.452 504	329.224 386	452.900 152	626.862 762
48	219.368 367	256.564.529	300.746 917	353.270 093	490.132 164	684.280 411
49	232.433 627	272.958 401	321.295 467	378.999 000	530.342 737	746.865 648
50	246.217 476	290.335 905	343.179 672	406.528 929	573.770 156	815.083 556
55	327.377 486	394.172 027	475.879 533	575.928 593	848.923 201	1 260.091 796
60	433.450 372	533.128 181	657.689 842	813.520 383	1 253.213 296	1 944.792 133
65	572.083 392	719.082 861	906.785 722	1 146.755 161	1 847.248 083	2 998.288 474
70	753.271 204	967.932 170	1 248.068 666	1 614.134 174	2 720.080 074	4 619.223 180
75	990.076 429	1 300.948 680	1 715.655 875	2 269.657 419	4 002.556 624	7 113.232 148
80	1 299.571 387	1 746.599 891	2 356.290 874	3 189.062 680	5 886.935 428	10 950.574 090
85	1 704.068 919	2 342.981 741	3 234.016 343	4 478.576 120	8 655.706 112	16 854.800 326
90	2 232.731 017	3 141.075 187	4 436.576 302	6 287.185 427	12 723.938 616	25 939.184 247
95	2 923.671 235	4 209.104 250	6 084.187 663	8 823.853 541	18 701.506 857	39 916.634 964
100	3 826.702 467	5 638.368 059	8 341.558 016	12 381.661 794	27 484.515 704	61 422.675 465

TABLE 8 (continued): $s_{\overline{n}|}\ i = \dfrac{(1+i)^n - 1}{i}$

n \ i	10 % 0.1	11 % 0.11	12 % 0.12	13 % 0.13	14 % 0.14	15 % 0.15
1	1.000 000	1.000 000	1.000 000	1.000 000	1.000 000	1.000 000
2	2.100 000	2.110 000	2.120 000	2.130 000	2.140 000	2.150 000
3	3.310 000	3.342 100	3.374 400	3.406 900	3.439 600	3.472 500
4	4.641 000	4.709 731	4.779 328	4.849 797	4.921 144	4.993 375
5	6.105 100	6.227 801	6.352 847	6.480 271	6.610 104	6.742 381
6	7.715 610	7.912 860	8.115 189	8.322 706	8.535 519	8.753 738
7	9.487 171	9.783 274	10.089 012	10.404 658	10.730 491	11.066 799
8	11.435 888	11.859 434	12.299 693	12.757 263	13.232 760	13.726 819
9	13.579 477	14.163 972	14.775 656	15.415 707	16.085 347	16.785 842
10	15.937 425	16.722 009	17.548 735	18.419 749	19.337 295	20.303 718
11	18.531 167	19.561 430	20.654 583	21.814 317	23.044 516	24.349 276
12	21.384 284	22.713 187	24.133 133	25.650 178	27.270 749	29.001 667
13	24.522 712	26.211 638	28.029 109	29.984 701	32.088 654	34.351 917
14	27.974 983	30.094 918	32.392 602	34.882 712	37.581 065	40.504 705
15	31.772 482	34.405 359	37.279 715	40.417 464	43.842 414	47.580 411
16	35.949 730	39.189 948	42.753 280	46.671 735	50.980 352	55.717 472
17	40.544 703	44.500 843	48.883 674	53.739 060	59.117 601	65.075 093
18	45.599 173	50.395 936	55.749 715	61.725 138	68.394 066	75.836 357
19	51.159 090	56.939 488	63.439 681	70.749 406	78.969 235	88.211 811
20	57.274 999	64.202 832	72.052 442	80.946 829	91.024 928	102.443 583
21	64.002 499	72.265 144	81.698 736	92.469 917	104.768 418	118.810 120
22	71.402 749	81.214 309	92.502 584	105.491 006	120.435 996	137.631 638
23	79.543 024	91.147 884	104.602 894	120.204 837	138.297 035	159.276 384
24	88.497 327	102.174 151	118.155 241	136.831 465	158.658 620	184.167 841
25	98.347 059	114.413 307	133.333 870	155.619 556	181.870 827	212.793 017
26	109.181 765	127.998 771	150.333 934	176.850 098	208.332 743	245.711 970
27	121.099 942	143.078 636	169.374 007	200.840 611	238.499 327	283.568 766
28	134.209 936	159.817 286	190.698 887	227.949 890	272.889 233	327.104 080
29	148.630 930	178.397 187	214.582 754	258.583 376	312.093 725	377.169 693
30	164.494 023	199.020 878	241.332 684	293.199 215	356.786 847	434.745 146
31	181.943 425	221.913 174	271.292 606	332.315 113	407.737 006	500.956 918
32	201.137 767	247.323 624	304.847 719	376.516 078	465.820 186	577.100 456
33	222.251 544	275.529 222	342.429 446	426.463 168	532.035 012	664.665 524
34	245.476 699	306.837 437	384.520 979	482.903 380	607.519 914	765.365 353
35	271.024 368	341.589 555	431.663 496	546.680 819	693.572 702	881.170 156
36	299.126 805	380.164 406	484.463 116	618.749 325	791.672 881	1 014.345 680
37	330.039 486	422.982 490	543.598 690	700.186 738	903.507 084	1 167.497 532
38	364.043 434	470.510 564	609.830 533	792.211 014	1 030.998 076	1 343.622 161
39	401.447 778	523.266 726	684.010 197	896.198 445	1 176.337 806	1 546.165 485
40	442.592 556	581.826 066	767.091 420	1 013.704 243	1 342.025 099	1 779.090 308
41	487.851 811	646.826 934	860.142 391	1 146.485 795	1 530.908 613	2 046.953 854
42	537.636 992	718.977 896	964.359 478	1 296.528 948	1 746.235 819	2 354.996 933
43	592.400 692	799.065 465	1 081.082 615	1 466.077 712	1 991.708 833	2 709.246 473
44	652.640 761	887.962 666	1 211.812 529	1 657.667 814	2 271.548 070	3 116.633 433
45	718.904 837	986.638 559	1 358.230 032	1 874.164 630	2 590.564 800	3 585.128 460
46	791.795 321	1 096.168 801	1 522.217 636	2 118.806 032	2 954.243 872	4 123.897 729
47	871.974 853	1 217.747 369	1 705.883 752	2 395.250 816	3 368.838 014	4 743.482 388
48	960.172 338	1 352.699 580	1 911.589 803	2 707.633 422	3 841.475 336	5 456.004 746
49	1 057.189 572	1 502.496 533	2 141.980 579	3 060.625 767	4 380.281 883	6 275.405 458
50	1 163.908 529	1 668.771 152	2 400.018 249	3 459.507 117	4 994.521 346	7 217.716 277
55	1 880.591 425	2 818.204 240	4 236.005 047	6 380.397 885	9 623.134 336	14 524.147 893
60	3 034.816 395	4 755.065 839	7 471.641 112	11 761.949 792	18 535.133 283	29 219.991 638
65	4 893.707 253	8 018.790 272	13 173.937 422	21 677.110 345	35 694.426 015	58 778.582 580
70	7 887.469 568	13 518.355 744	23 223.331 897	39 945.150 956	68 733.178 463	118 231.466 926
75	12 708.953 714	22 785.443 391	40 933.798 673	73 602.831 635	132 346.474 212	237 812.453 171
80	20 474.002 146	38 401.025 004	72 145.692 501	135 614.926 571	254 828.441 480	478 332.529 343
85	32 979.690 296	64 714.188 149	127 151.714 005	249 868.191 823	490 657.007 341	962 104.313 290
90	53 120.226 118	109 053.398 293	224 091.118 528	460 372.427 073	944 724.766 995	1 935 142.168 042
95	85 556.760 466	183 767.545 936	394 931.471 864	848 212.835 490	1 818 993.452 831	3 892 268.850 907
100	137 796.123 398	309 665.229 724	696 010.547 721	1 562 783.647 911	3 502 323.129 475	7 828 749.671 335

STATISTICAL AND INTEREST TABLES

TABLE 9: $a_{\overline{n}|}\ i = \dfrac{1-(1+i)^{-n}}{i}$

n \ i	½ % 0.005	¾ % 0.0075	1 % 0.01	1¼ % 0.0125	1½ % 0.015	2 % 0.02
1	0.995 025	0.992 556	0.990 099	0.987 654	0.985 222	0.980 392
2	1.985 099	1.977 723	1.970 395	1.963 115	1.955 883	1.941 561
3	2.970 248	2.955 556	2.940 985	2.926 534	2.912 200	2.883 883
4	3.950 496	3.926 110	3.901 966	3.878 058	3.854 385	3.807 729
5	4.925 866	4.889 440	4.853 431	4.817 835	4.782 645	4.713 460
6	5.896 384	5.845 598	5.795 476	5.746 010	5.697 187	5.601 431
7	6.862 074	6.794 638	6.728 195	6.662 726	6.598 214	6.471 991
8	7.822 959	7.736 613	7.651 678	7.568 124	7.485 925	7.325 481
9	8.779 064	8.671 576	8.566 018	8.462 345	8.360 517	8.162 237
10	9.730 412	9.599 580	9.471 305	9.345 526	9.222 185	8.982 585
11	10.677 027	10.520 675	10.367 628	10.217 803	10.071 118	9.786 848
12	11.618 932	11.434 913	11.255 077	11.079 312	10.907 505	10.575 341
13	12.556 151	12.342 345	12.133 740	11.930 185	11.731 532	11.348 374
14	13.488 708	13.243 022	13.003 703	12.770 553	12.543 382	12.106 249
15	14.416 625	14.136 995	13.865 053	13.600 546	13.343 233	12.849 264
16	15.339 925	15.024 313	14.717 874	14.420 292	14.131 264	13.577 709
17	16.258 632	15.905 025	15.562 251	15.229 918	14.907 649	14.291 872
18	17.172 768	16.779 181	16.398 269	16.029 549	15.672 561	14.992 031
19	18.082 356	17.646 830	17.226 008	16.819 308	16.426 168	15.678 462
20	18.987 419	18.508 020	18.045 553	17.599 316	17.168 639	16.351 433
21	19.887 979	19.362 799	18.856 983	18.369 695	17.900 137	17.011 209
22	20.784 059	20.211 215	19.660 379	19.130 563	18.620 824	17.658 048
23	21.675 681	21.053 315	20.455 821	19.882 037	19.330 861	18.292 204
24	22.562 866	21.889 146	21.243 387	20.624 235	20.030 405	18.913 926
25	23.445 638	22.718 755	22.023 156	21.357 269	20.719 611	19.523 456
26	24.324 018	23.542 189	22.795 204	22.081 253	21.398 632	20.121 036
27	25.198 028	24.359 493	23.559 608	22.796 299	22.067 617	20.706 898
28	26.067 689	25.170 713	24.316 443	23.502 518	22.726 717	21.281 272
29	26.933 024	25.975 893	25.065 785	24.200 018	23.376 076	21.844 385
30	27.794 054	26.775 080	25.807 708	24.888 906	24.015 838	22.396 456
31	28.650 800	27.568 318	26.542 285	25.569 290	24.646 146	22.937 702
32	29.503 284	28.355 650	27.269 589	26.241 274	25.267 139	23.468 335
33	30.351 526	29.137 122	27.989 693	26.904 962	25.878 954	23.988 564
34	31.195 548	29.912 776	28.702 666	27.560 456	26.481 728	24.498 592
35	32.035 371	30.682 656	29.408 580	28.207 858	27.075 595	24.998 619
36	32.871 016	31.446 805	30.107 505	28.847 267	27.660 684	25.488 842
37	33.702 504	32.205 266	30.799 510	29.478 783	28.237 127	25.969 453
38	34.529 854	32.958 080	31.484 663	30.102 501	28.805 052	26.440 641
39	35.353 089	33.705 290	32.163 033	30.718 520	29.364 583	26.902 589
40	36.172 228	34.446 938	32.834 686	31.326 933	29.915 845	27.355 479
41	36.987 291	35.183 065	33.499 689	31.927 835	30.458 961	27.799 489
42	37.798 300	35.913 713	34.158 108	32.521 319	30.994 050	28.234 794
43	38.605 274	36.638 921	34.810 008	33.107 475	31.521 232	28.661 562
44	39.408 232	37.358 730	35.455 454	33.686 395	32.040 622	29.079 963
45	40.207 196	38.073 181	36.094 508	34.258 168	32.552 337	29.490 160
46	41.002 185	38.782 314	36.727 236	34.822 882	33.056 490	29.892 314
47	41.793 219	39.486 168	37.353 699	35.380 624	33.553 192	30.286 582
48	42.580 318	40.184 782	37.973 959	35.931 481	34.042 554	30.673 120
49	43.363 500	40.878 195	38.588 079	36.475 537	34.524 683	31.052 078
50	44.142 786	41.566 447	39.196 118	37.012 876	34.999 688	31.423 606
55	47.981 445	44.931 612	42.147 192	39.601 687	37.271 467	33.174 788
60	51.725 561	48.173 374	44.955 038	42.034 592	39.380 269	34.760 887
65	55.377 461	51.296 257	47.626 608	44.320 980	41.337 786	36.197 466
70	58.939 418	54.304 622	50.168 514	46.469 676	43.154 872	37.498 619
75	62.413 645	57.202 668	52.587 051	48.488 970	44.841 600	38.677 114
80	65.802 305	59.994 440	54.888 206	50.386 657	46.407 323	39.744 514
85	69.107 505	62.683 836	57.077 676	52.170 060	47.860 722	40.711 290
90	72.331 300	65.274 609	59.160 881	53.846 060	49.209 855	41.586 929
95	75.475 694	67.770 377	61.142 980	55.421 127	50.462 201	42.380 023
100	78.542 645	70.174 623	63.028 879	56.901 339	51.624 704	43.098 352

TABLE 9 (continued): $a_{\overline{n}|}\ i = \dfrac{1-(1+i)^{-n}}{i}$

n \ i	2½ % 0.025	3 % 0.03	3½ % 0.035	4 % 0.04	4½ % 0.045	5 % 0.05
1	0.975 610	0.970 874	0.966 184	0.961 538	0.956 938	0.952 381
2	1.927 424	1.913 470	1.899 694	1.886 095	1.872 668	1.859 410
3	2.856 024	2.828 611	2.801 637	2.775 091	2.748 964	2.723 248
4	3.761 974	3.717 098	3.673 079	3.629 895	3.587 526	3.545 951
5	4.645 828	4.579 707	4.515 052	4.451 822	4.389 977	4.329 477
6	5.508 125	5.417 191	5.328 553	5.242 137	5.157 872	5.075 692
7	6.349 391	6.230 283	6.114 544	6.002 055	5.892 701	5.786 373
8	7.170 137	7.019 692	6.873 956	6.732 745	6.595 886	6.463 213
9	7.970 866	7.786 109	7.607 687	7.435 332	7.268 790	7.107 822
10	8.752 064	8.530 203	8.316 605	8.110 896	7.912 718	7.721 735
11	9.514 209	9.252 624	9.001 551	8.760 477	8.528 917	8.306 414
12	10.257 765	9.954 004	9.663 334	9.385 074	9.118 581	8.863 252
13	10.983 185	10.634 955	10.302 738	9.985 648	9.682 852	9.393 573
14	11.690 912	11.296 073	10.920 520	10.563 123	10.222 825	9.898 641
15	12.381 378	11.937 935	11.517 411	11.118 387	10.739 546	10.379 658
16	13.055 003	12.561 102	12.094 117	11.652 296	11.234 015	10.837 770
17	13.712 198	13.166 118	12.651 321	12.165 669	11.707 191	11.274 066
18	14.353 364	13.753 513	13.189 682	12.659 297	12.159 992	11.689 587
19	14.978 891	14.323 799	13.709 837	13.133 939	12.593 294	12.085 321
20	15.589 162	14.877 475	14.212 403	13.590 326	13.007 936	12.462 210
21	16.184 549	15.416 024	14.697 974	14.029 160	13.404 724	12.821 153
22	16.765 413	15.936 917	15.167 125	14.451 115	13.784 425	13.163 003
23	17.332 110	16.443 608	15.620 410	14.856 842	14.147 775	13.488 574
24	17.884 986	16.935 542	16.058 368	15.246 963	14.495 478	13.798 642
25	18.424 376	17.413 148	16.481 515	15.622 080	14.828 209	14.093 945
26	18.950 611	17.876 842	16.890 352	15.982 769	15.146 611	14.375 185
27	19.464 011	18.327 031	17.285 365	16.329 586	15.451 303	14.643 034
28	19.964 889	18.764 108	17.667 019	16.663 063	15.742 874	14.898 127
29	20.453 550	19.188 455	18.035 767	16.983 715	16.021 889	15.141 074
30	20.930 293	19.600 441	18.392 045	17.292 033	16.288 889	15.372 451
31	21.395 407	20.000 428	18.736 276	17.588 494	16.544 391	15.592 811
32	21.849 178	20.388 766	19.068 865	17.873 551	16.788 891	15.802 677
33	22.291 881	20.765 792	19.390 208	18.147 646	17.022 862	16.002 549
34	22.723 786	21.131 837	19.700 684	18.411 198	17.246 758	16.192 904
35	23.145 157	21.487 220	20.000 661	18.664 613	17.461 012	16.374 194
36	23.556 251	21.832 252	20.290 494	18.908 282	17.666 041	16.546 852
37	23.957 318	22.167 235	20.570 525	19.142 579	17.862 240	16.711 287
38	24.348 603	22.492 462	20.841 087	19.367 864	18.049 990	16.867 893
39	24.730 344	22.808 215	21.102 500	19.584 485	18.229 656	17.017 041
40	25.102 775	23.114 772	21.355 072	19.792 774	18.401 584	17.159 086
41	25.466 122	23.412 400	21.599 104	19.993 052	18.566 109	17.294 368
42	25.820 607	23.701 359	21.834 883	20.185 627	18.723 550	17.423 208
43	26.166 446	23.981 902	22.062 689	20.370 795	18.874 210	17.545 912
44	26.503 849	24.254 274	22.282 791	20.548 841	19.018 383	17.662 773
45	26.833 024	24.518 713	22.495 450	20.720 040	19.156 347	17.774 070
46	27.154 170	24.775 449	22.700 918	20.884 654	19.288 371	17.880 066
47	27.467 483	25.024 708	22.899 438	21.042 936	19.414 709	17.981 016
48	27.773 154	25.266 707	23.091 244	21.195 131	19.535 607	18.077 158
49	28.071 369	25.501 657	23.276 564	21.341 472	19.651 298	18.168 722
50	28.362 312	25.729 764	23.455 618	21.482 185	19.762 008	18.255 925
55	29.713 979	26.774 428	24.264 053	22.108 612	20.248 021	18.633 472
60	30.908 656	27.675 564	24.944 734	22.623 490	20.638 022	18.929 290
65	31.964 577	28.452 892	25.517 849	23.046 682	20.950 979	19.161 070
70	32.897 857	29.123 421	26.000 397	23.394 515	21.202 112	19.342 677
75	33.722 740	29.701 826	26.406 689	23.680 408	21.403 634	19.484 970
80	34.451 817	30.200 763	26.748 776	23.915 392	21.565 345	19.596 460
85	35.096 215	30.631 151	27.036 804	24.108 531	21.695 110	19.683 816
90	35.665 768	31.002 407	27.279 316	24.267 278	21.799 241	19.752 262
95	36.169 171	31.322 656	27.483 504	24.397 756	21.882 800	19.805 891
100	36.614 105	31.598 905	27.655 425	24.504 999	21.949 853	19.847 910

STATISTICAL AND INTEREST TABLES

TABLE 9 (continued): $a_{\overline{n}|}\ i = \dfrac{1-(1+i)^{-n}}{i}$

i $\diagdown$ n	5½ % 0.055	6 % 0.06	6½ % 0.065	7 % 0.07	8 % 0.08	9 % 0.09
1	0.947 867	0.943 396	0.938 967	0.934 579	0.925 926	0.917 431
2	1.846 320	1.833 393	1.820 626	1.808 018	1.783 265	1.759 111
3	2.697 933	2.673 012	2.648 476	2.624 316	2.577 097	2.531 295
4	3.505 150	3.465 106	3.425 799	3.387 211	3.312 127	3.239 720
5	4.270 284	4.212 364	4.155 679	4.100 197	3.992 710	3.889 651
6	4.995 530	4.917 324	4.841 014	4.766 540	4.622 880	4.485 919
7	5.682 967	5.582 381	5.484 520	5.389 289	5.206 370	5.032 953
8	6.334 566	6.209 794	6.088 751	5.971 299	5.746 639	5.534 819
9	6.952 195	6.801 692	6.656 104	6.515 232	6.246 888	5.995 247
10	7.537 626	7.360 087	7.188 830	7.023 582	6.710 081	6.417 658
11	8.092 536	7.886 875	7.689 042	7.498 674	7.138 964	6.805 191
12	8.618 518	8.383 844	8.158 725	7.942 686	7.536 078	7.160 725
13	9.117 079	8.852 683	8.599 742	8.357 651	7.903 776	7.486 904
14	9.589 648	9.294 984	9.013 842	8.745 468	8.244 237	7.786 150
15	10.037 581	9.712 249	9.402 669	9.107 914	8.559 479	8.060 688
16	10.462 162	10.105 895	9.767 764	9.446 649	8.851 369	8.312 558
17	10.864 609	10.477 260	10.110 577	9.763 223	9.121 638	8.543 631
18	11.246 074	10.827 603	10.432 466	10.059 087	9.371 887	8.755 625
19	11.607 654	11.158 116	10.734 710	10.335 595	9.603 599	8.950 115
20	11.950 382	11.469 921	11.018 507	10.594 014	9.818 147	9.128 546
21	12.275 244	11.764 077	11.284 983	10.835 527	10.016 803	9.292 244
22	12.583 170	12.041 582	11.535 196	11.061 240	10.200 744	9.442 425
23	12.875 042	12.303 379	11.770 137	11.272 187	10.371 059	9.580 207
24	13.151 699	12.550 358	11.990 739	11.469 334	10.528 758	9.706 612
25	13.413 933	12.783 356	12.197 877	11.653 583	10.674 776	9.822 580
26	13.662 495	13.003 166	12.392 373	11.825 779	10.809 978	9.928 972
27	13.898 100	13.210 534	12.574 998	11.986 709	10.935 165	10.026 580
28	14.121 422	13.406 164	12.746 477	12.137 111	11.051 078	10.116 128
29	14.333 101	13.590 721	12.907 490	12.277 674	11.158 406	10.198 283
30	14.533 745	13.764 831	13.058 676	12.409 041	11.257 783	10.273 654
31	14.723 929	13.929 086	13.200 635	12.531 814	11.349 799	10.342 802
32	14.904 198	14.084 043	13.333 929	12.646 555	11.434 999	10.406 240
33	15.075 069	14.230 230	13.459 088	12.753 790	11.513 888	10.464 441
34	15.237 033	14.368 141	13.576 609	12.854 009	11.586 934	10.517 835
35	15.390 552	14.498 246	13.686 957	12.947 672	11.654 568	10.566 821
36	15.536 068	14.620 987	13.790 570	13.035 208	11.717 193	10.611 763
37	15.673 999	14.736 780	13.887 859	13.117 017	11.775 179	10.652 993
38	15.804 738	14.846 019	13.979 210	13.193 473	11.828 869	10.690 820
39	15.928 662	14.949 075	14.064 986	13.264 928	11.878 582	10.725 523
40	16.046 125	15.046 297	14.145 527	13.331 709	11.924 613	10.757 360
41	16.157 464	15.138 016	14.221 152	13.394 120	11.967 235	10.786 569
42	16.262 999	15.224 543	14.292 161	13.452 449	12.006 699	10.813 366
43	16.363 032	15.306 173	14.358 837	13.506 962	12.043 240	10.837 950
44	16.457 851	15.383 182	14.421 443	13.557 908	12.077 074	10.860 505
45	16.547 726	15.455 832	14.480 228	13.605 522	12.108 402	10.881 197
46	16.632 915	15.524 370	14.535 426	13.650 020	12.137 409	10.900 181
47	16.713 664	15.589 028	14.587 254	13.691 608	12.164 267	10.917 597
48	16.790 203	15.650 027	14.635 919	13.730 474	12.189 136	10.933 575
49	16.862 751	15.707 572	14.681 615	13.766 799	12.212 163	10.948 234
50	16.931 518	15.761 861	14.724 521	13.800 746	12.233 485	10.961 683
55	17.225 170	15.990 543	14.902 825	13.939 939	12.318 614	11.013 993
60	17.449 854	16.161 428	15.032 966	14.039 181	12.376 552	11.047 991
65	17.621 767	16.289 123	15.127 953	14.109 940	12.415 983	11.070 087
70	17.753 304	16.384 544	15.197 282	14.160 389	12.442 820	11.084 449
75	17.853 947	16.455 848	15.247 885	14.196 359	12.461 084	11.093 782
80	17.930 953	16.509 131	15.284 818	14.222 005	12.473 514	11.099 849
85	17.989 873	16.548 947	15.311 775	14.240 297	12.481 974	11.103 791
90	18.034 954	16.578 699	15.331 451	14.253 328	12.487 732	11.106 354
95	18.069 447	16.600 932	15.345 812	14.262 623	12.491 651	11.108 019
100	18.095 839	16.617 546	15.356 293	14.269 251	12.494 318	11.109 102

TABLE 9 (continued): $a_{\overline{n}|}\ i = \dfrac{1-(1+i)^{-n}}{i}$

i \ n	10 % 0.1	11 % 0.11	12 % 0.12	13 % 0.13	14 % 0.14	15 % 0.15
1	0.909 091	0.900 901	0.892 857	0.884 956	0.877 193	0.869 565
2	1.735 537	1.712 523	1.690 051	1.668 102	1.646 661	1.625 709
3	2.486 852	2.443 715	2.401 831	2.361 153	2.321 632	2.283 225
4	3.169 865	3.102 446	3.037 349	2.974 471	2.913 712	2.854 978
5	3.790 787	3.695 897	3.604 776	3.517 231	3.433 081	3.352 155
6	4.355 261	4.230 538	4.111 407	3.997 550	3.888 668	3.784 483
7	4.868 419	4.712 196	4.563 757	4.422 610	4.288 305	4.160 420
8	5.334 926	5.146 123	4.967 640	4.798 770	4.638 864	4.487 322
9	5.759 024	5.537 048	5.328 250	5.131 655	4.946 372	4.771 584
10	6.144 567	5.889 232	5.650 223	5.426 243	5.216 116	5.018 769
11	6.495 061	6.206 515	5.937 699	5.686 941	5.452 733	5.233 712
12	6.813 692	6.492 356	6.194 374	5.917 647	5.660 292	5.420 619
13	7.103 356	6.749 870	6.423 548	6.121 812	5.842 362	5.583 147
14	7.366 687	6.981 865	6.628 168	6.302 488	6.002 072	5.724 476
15	7.606 080	7.190 870	6.810 864	6.462 379	6.142 168	5.847 370
16	7.823 709	7.379 162	6.973 986	6.603 875	6.265 060	5.954 235
17	8.021 553	7.548 794	7.119 630	6.729 093	6.372 859	6.047 161
18	8.201 412	7.701 617	7.249 670	6.839 905	6.467 420	6.127 966
19	8.364 920	7.839 294	7.365 777	6.937 969	6.550 369	6.198 231
20	8.513 564	7.963 328	7.469 444	7.024 752	6.623 131	6.259 331
21	8.648 694	8.075 070	7.562 003	7.101 550	6.686 957	6.312 462
22	8.771 540	8.175 739	7.644 646	7.169 513	6.742 944	6.358 663
23	8.883 218	8.266 432	7.718 434	7.229 658	6.792 056	6.398 837
24	8.984 744	8.348 137	7.784 316	7.282 883	6.835 137	6.433 771
25	9.077 040	8.421 745	7.843 139	7.329 985	6.872 927	6.464 149
26	9.160 945	8.488 058	7.895 660	7.371 668	6.906 077	6.490 564
27	9.237 223	8.547 800	7.942 554	7.408 556	6.935 155	6.513 534
28	9.306 567	8.601 622	7.984 423	7.441 200	6.960 662	6.533 508
29	9.369 606	8.650 110	8.021 806	7.470 088	6.983 037	6.550 877
30	9.426 914	8.693 793	8.055 184	7.495 653	7.002 664	6.565 980
31	9.479 013	8.733 146	8.084 986	7.518 277	7.019 881	6.579 113
32	9.526 376	8.768 600	8.111 594	7.538 299	7.034 983	6.590 533
33	9.569 432	8.800 541	8.135 352	7.556 016	7.048 231	6.600 463
34	9.608 575	8.829 316	8.156 564	7.571 696	7.059 852	6.609 099
35	9.644 159	8.855 240	8.175 504	7.585 572	7.070 045	6.616 607
36	9.676 508	8.878 594	8.192 414	7.597 851	7.078 987	6.623 137
37	9.705 917	8.899 635	8.207 513	7.608 718	7.086 831	6.628 815
38	9.732 651	8.918 590	8.220 993	7.618 334	7.093 711	6.633 752
39	9.756 956	8.935 666	8.233 030	7.626 844	7.099 747	6.638 045
40	9.779 051	8.951 051	8.243 777	7.634 376	7.105 041	6.641 778
41	9.799 137	8.964 911	8.253 372	7.641 040	7.109 685	6.645 025
42	9.817 397	8.977 397	8.261 939	7.646 938	7.113 759	6.647 848
43	9.833 998	8.988 646	8.269 589	7.652 158	7.117 332	6.650 302
44	9.849 089	8.998 780	8.276 418	7.656 777	7.120 467	6.652 437
45	9.862 808	9.007 910	8.282 516	7.660 864	7.123 217	6.654 293
46	9.875 280	9.016 135	8.287 961	7.664 482	7.125 629	6.655 907
47	9.886 618	9.023 545	8.292 822	7.667 683	7.127 744	6.657 310
48	9.896 926	9.030 221	8.297 163	7.670 516	7.129 600	6.658 531
49	9.906 296	9.036 235	8.301 038	7.673 023	7.131 228	6.659 592
50	9.914 814	9.041 653	8.304 498	7.675 242	7.132 656	6.660 515
55	9.947 106	9.061 678	8.316 972	7.683 045	7.137 559	6.663 608
60	9.967 157	9.073 562	8.324 049	7.687 280	7.140 106	6.665 146
65	9.979 607	9.080 614	8.328 065	7.689 579	7.141 428	6.665 911
70	9.987 338	9.084 800	8.330 344	7.690 827	7.142 115	6.666 291
75	9.992 138	9.087 283	8.331 637	7.691 504	7.142 472	6.666 480
80	9.995 118	9.088 757	8.332 371	7.691 871	7.142 657	6.666 574
85	9.996 969	9.089 632	8.332 787	7.692 071	7.142 753	6.666 620
90	9.998 118	9.090 151	8.333 023	7.692 179	7.142 803	6.666 644
95	9.998 831	9.090 459	8.333 157	7.692 238	7.142 829	6.666 655
100	9.999 274	9.090 642	8.333 234	7.692 270	7.142 843	6.666 661